BOTANICAL RESEARCH AND PRACTICES

MOSS

CLASSIFICATION, DEVELOPMENT AND GROWTH AND FUNCTIONAL ROLE IN ECOSYSTEMS

Botanical Research and Practices

Additional books in this series can be found on Nova's website under the Series tab.

Additional E-books in this series can be found on Nova's website under the E-book tab.

BOTANICAL RESEARCH AND PRACTICES

MOSS

CLASSIFICATION, DEVELOPMENT AND GROWTH AND FUNCTIONAL ROLE IN ECOSYSTEMS

JORMA MOHAMED
EDITOR

nova publishers
New York

For permission to use material from this book please contact us:
Telephone 631-231-7269; Fax 631-231-8175
Web Site: http://www.novapublishers.com

Additional color graphics may be available in the e-book version of this book.

Library of Congress Cataloging-in-Publication Data

ISBN: 978-1-63117-396-7

Published by Nova Science Publishers, Inc. † New York

CONTENTS

PREFACE

This book examines microbial communities in terrestrial moss in further detail. It also discusses terrestrial mosses as trace element bioindicators; moss bioreactors; the ecological diversity and application for agricultural microbiology; small scale area investigation for heavy metals air pollution; biomonitoring of atmospheric pollution using mosses; the impact of forest decline on fine-scale vertical distribution of epiphytic bryophytes; and the use of mosses in criminal investigations.

Chapter 1 – Terrestrial moss communities are widely distributed among the Earth's major biomes, including equatorial to polar latitudes. Because of their high primary productivity, relatively dense vegetative structure, and alteration of the substratum through deposition of organic matter, moss communities often support rich and diversified microbial communities. This review encompasses soil-dwelling moss communities in low and high latitude locales, with special attention to moss-based microbial community structure (taxonomic composition and organization), function (microbial roles and interactions), and ecology (especially the dynamic interaction of microbes with mosses and their role in the carbon biogeochemical cycle). Nitrogen-fixing cyanobacteria, in addition to heterotrophic microbes, are significant members of some moss microbial communities, and consequently attention is given to the biogeochemistry of N and C.

Chapter 2 – Terrestrial mosses have been used as bioindicators since the 1960s. The first, large-scale studies which reported the concentrations of selected trace elements (As, Cd, Cr, Cu, Fe, Hg, Ni, Pb, V, Zn) were conducted in the Scandinavia in 1985 (Poikolainen et al., 2004). Since then many European scientists have employed the moss technique to solve various issues: (i) to document atmospheric chemistry, (ii) to map ecogeochemical changes in the environment, (iii) to pinpoint the major sources of atmospheric pollution, and (iv) to find contrast between background and "hot spot" element concentrations in polluted areas.

Terrestrial mosses have found practical applications in biomonitoring studies because of their morphological and physiological properties. The lack of cuticle and root system causes that their cell walls are easily penetrated by metal ions. This enables absorption of elements and particulates from rainfall, snowmelt and dry deposition. They also have a high cation exchange capacity to retain and transport many elements. Mosses are widespread, they grow in various habitats, and their annual growth increments are easily identified.

The concentrations of trace elements in moss tissues are dependent on individual moss species and their morphology, as well as bedrock lithology and many other environmental factors, i.e., topography, climate, hydrology, and edaphic (soil-related) conditions. Local topographic, climatic and edaphic variables are generally correlated with the geologic setting of study area. Different moss species show distinct variations in the concentrations of elements and some organics. This is induced primarily by diverse assimilation abilities. These facts put some constraints on the selection of a specific moss species for biomonitoring purposes. Several species (*Hylocomium splendens*, *Pleurozium schreberi*, *Scleropodium purum*, *Hypnum cupressiforme*) have successfully been used in biomonitoring programs, but all of these species must have acceptable sampling density, which is crucial to identify the species-dependent variability.

It should be stressed that the moss technique has many advantages. It is easier and cheaper than the conventional analysis of precipitation, which is based on a large number of sample collectors. This technique also allows us to monitor several sites at a low cost. Moreover, the element concentrations in moss tissues are high enough to make the analysis much simpler.

Chapter 3 – The use of moss bioreactors for the production of proteins and peptides is a recently incorporated heterologous system that was developed as an alternative for the expression of recombinant molecules that were not well obtained in mammals, insect and/or yeast systems. Moss is classified into plant kingdom, but it contains a very small genome of 511 Mbps, which facilitates its manipulation and gene transformation. Moreover, it presents less risk of contaminations with human and animal pathogens, when compared with expression systems using bacterial and mammalian cells. Nowadays, the moss species most used as a bioreactor is *Physcomitrella patens*. It was earlier used for studies on plant evolution, development and physiology. Since the last 30 years, *P. patens* started being evaluated as a new option for the production of recombinant proteins. The advantages of moss over other higher plant species is the enhanced rates of homologous recombination, giving straight-forward reverse genetics protocols by gene targeting. Furthermore, it lacks a bias for codon usage, making it less complicated to adapt foreign genes in its genome for protein expression. The cultivation of *P. patens* is also relatively simple and of low cost, when compared to higher plant species. It can grow under a wide range of pH and temperature, using light and carbon dioxide as its only energy and carbon sources. Until now, most proteins expressed in moss systems are from the human genome, with the purpose of producing biopharmaceutical molecules. Hence, it has been described the use of moss bioreactors for the production of monoclonal antibodies, vaccines-to-be molecules and other human proteins and peptides. Therefore, this report aims to describe the most recent advances on the use of moss bioreactor for the expression of proteins targeted as future biopharmaceutical molecules.

Chapter 4 – *Sphagnum* mosses serve as a unique habitat for microorganisms, which play an important role both for the host plants and the peatland ecosystems. The aim of the present study was to isolate endophytic bacteria from the tissues of *Sphagnum* mosses and to screen them for strains promising for further application in agricultural microbiology. About 50 samples of *Sphagnum fallax* (H. Klinggr.) H. Klinggr. and *Sphagnum magellanicum* Brid. were collected in the Austrian Alps, Lenindgrad Region and West Siberia in 2009–2012. Endophytic bacteria were detected inside the moss plants using fluorescent in situ hybridization (FISH) followed by confocal laser scanning microscopy (CLSM). Altogether, more than 400 isolates were obtained by cultivation on the nutrient media. Examination of the isolates for the antagonistic activity revealed that more than 50% of them could suppress the growth of phytopathogenic and toxigenic fungi. More than 30% of isolates showed some antagonistic activity against microbial phytopathogens. The isolated strains could colonize crops and promote their growth. Molecular-genetic identification coupled with physiological/ biochemical characterization showed that the dominant endophytic groups belonged to the genera *Burkholderia*, *Pseudomonas*, *Flavobacterium*, *Serratia* and *Collimonas*. The isolated endophytes were shown to be promising objects for the development of effective growth promoting and protective microbiological preparations to be used in agriculture.

Chapter 5 – Application of several moss species for monitoring of the anthropogenic impact on heavy metals air pollution in a small scale area was studied. Mosses were reviewed for their potential to reflect metal air pollution. The attention was focused on their quantification ability, underlying the metal accumulation in the moss plant tissue. Potential "hot spots" were selected in areas of lead and zinc mines (case study at Zletovo mine and Sasa mine), copper mine (Bučim mine), ferronickel smelter plant (FENI industry) and the abandoned As-Sb-Tl mine at the Allchar area, as the main metal pollution sources in the Republic of Macedonia in the period between 2010-2012. There is continuous distribution of dust from ore, flotation tailings and slag surface. This results in air-introduction and deposition of higher contents of certain metals. Several moss species (*Hypnum cupressiforme*, *Campothecium lutescens, Scleropodium purum* and *Homolothecium sericium*) were used as plant sampling media. Determination of chemical elements was conducted by using both instrumental techniques: atomic emission spectrometry with inductively coupled plasma (ICP-AES) and mass spectrometry with inductively coupled plasma (ICP-MS). Combination of multivariate techniques (PCA, FA and CA) was applied for data processing and identification of elements associated with lithogenic or anthropogenic origin. Spatial distribution maps were created for determination and localization of narrower areas with higher contents of certain anthropogenic elements. In this way, influences of selected human activities in local (small scale) air pollution cases can be determined. Summarized data reveal real quantification of elements distribution not only in order to determine the hazardous elements distribution, but also present complete characterization of elements deposition in mines/smelter plant environs.

Chapter 6 – Due to their peculiar morphological and physiological characteristics (lack of a protective cuticle and thick cell walls, numerous cell wall constituents with negatively charged groups, mineral nutrition obtained mainly from wet and dry deposition) mosses are extensively used as biomonitors of environmental contamination for a wide range of contaminants, i.e. radionuclides, heavy metals and toxic nitrogen, sulphur and organic compounds and also as reference organisms in the assessment of radiation exposure to non-human biota. Mosses usually show considerable longevity, which assists their use as long-term integrators of atmospheric deposition. The use of mosses as biomonitors of atmospheric contamination is reviewed in this chapter, with the special reference to terrestrial ecosystems in Serbia, including data on radionuclide and heavy metal content in mosses, their spatial patterns and species- and site-specific dependencies.

Chapter 7 – The vertical distribution of epiphytic bryophytes on tree trunks may serve as an indicator of diversity, and requires delineation. This study assessed changes in the vertical distribution of epiphytic bryophytes to determine how forest decline affects the diversity of these non-vascular plants.

Epiphytic bryophytes growing on the trunks of *Pinus jezoensis* var. *hondoensis* trees were examined along four 50-cm-high zones from ground level to a trunk height of 2 m in six forest patches at different states of forest decline (no decline, slight decline, severe decline, and serious decline). Overall, epiphyte species richness and trunk coverage decreased with the process of forest decline. Using two-way indicator species analysis and detrended correspondence analysis, epiphytic bryophyte flora diversity appeared to shift from higher trunk zones in healthy forests to lower trunk zones in heavily degraded forests. These changes might be explained by an increase in drought stress and light intensity caused by forest decline, which might have a greater effect on the composition of epiphytic bryophytes inhabiting the lower zones of trunks. Therefore, the epiphytic bryophytes occupying the lower trunk zones in forest patches where forest decline continues require careful monitoring and conservation. As the fine-scale distribution of epiphytic bryophytes may be easily surveyed without any special equipment or machines, these plants could be used to regularly monitor and evaluate the impact of forest decline on epiphytic bryophytes.

Chapter 8 – Forensic Botany is the application of botanic knowledge to criminal investigations. It is a relatively new science and it includes several subdisciplines, such as Palynology, Dendrochronology, Limnology, Systematic, Ecology and Molecular Biology. Plant material is everywhere and it is frequently found in association with forensic cases, especially outdoor. Nevertheless its potential applications are often unknown or underestimated by the investigators and prosecutors, so botanical traces are usually overlooked. Instead a systematic analysis of botanical evidence can provide useful information about manner and time of death, find out and dating clandestine graves, identify the primary crime scene and help discovering the offender. Botanical evidence is mainly used in forensic field to provide linkages between crime scenes, individuals (victims or suspects) and objects through morphological, anatomic and genetic study of pollens and spores, seeds, living or decayed leaves, plant fragments or algae. Mosses are particularly useful for this purpose. There are over 10.000 species of mosses worldwide; they are ubiquitous (even if they prefer humid and shadowy zones) but their growth substrates and environmental conditions are variable; so they represent good "biological indicators", more reliable than other plant organisms for geolocation. Mosses can grow directly associated to human remains or objects; moss fragments are resistant to decay and they can heavily attach to clothes or shoes, where they can be discovered and identified by genetic analysis even after a long time. Moss spores found in evidentiary materials or human remains might be used to establish in which season a crime occurred. In the last years the use of mosses for forensic purposes expanded; in recent studies the regular growth rate of some species of mosses was used in estimating the minimum post-mortem interval of skeletal remains. In this chapter the authors give an overlook on possible application of moss in Forensics.

In: Moss
Editor: Jorma Mohamed
ISBN: 978-1-63117-396-7

Chapter 1

MICROBIAL COMMUNITIES IN TERRESTRIAL MOSS (BRYOPHYTA): STRUCTURE, FUNCTION AND ECOLOGICAL SIGNIFICANCE

O. Roger Anderson*
Biology, Lamont-Doherty Earth Observatory of Columbia University, Palisades, NY, US

ABSTRACT

Terrestrial moss communities are widely distributed among the Earth's major biomes, including equatorial to polar latitudes. Because of their high primary productivity, relatively dense vegetative structure, and alteration of the substratum through deposition of organic matter, moss communities often support rich and diversified microbial communities. This review encompasses soil-dwelling moss communities in low and high latitude locales, with special attention to moss-based microbial community structure (taxonomic composition and organization), function (microbial roles and interactions), and ecology (especially the dynamic interaction of microbes with mosses and their role in the carbon biogeochemical cycle). Nitrogen-fixing cyanobacteria, in addition to heterotrophic microbes, are significant members of some moss microbial

* Corresponding author: E-mail: ora@LDEO.columbia.edu.

communities, and consequently attention is given to the biogeochemistry of N and C.

Introduction

Mosses (Bryophyta) are a diverse group of photosynthetic, nonvascular plants widely distributed among the Earth's major biomes from the equator to the poles, and are major primary producers in the Arctic tundra and high latitude boreal forests encompassing millions of square kilometers globally (e.g., Anderson, 2010a; O'Neill, 2000; Walker and Walker, 1996). Mosses are recognized as important pioneering plants on barren, potentially productive terrestrial sites (e.g., Taylor, 1920). Moreover, due to their primary production, they provide rich organic residues to support other life forms, including microbial communities in early ecological successions (e.g., Clements, 1916; Rayburn et al., 1982; Wynn-Williams 1993) and during soil rehabilitation (Bowker, 2007). However, their role in microbial synecology is often overlooked, especially in relation to the global biogeochemical C-cycle. In some cases, the dynamic interaction among microbial species and mosses is limited to one or a few microbial groups, such as bacteria, fungi, or selected groups of protists, and less attention is given to the complex ways more composite suites of microbes relate to their moss habitats. This review highlights some of the current knowledge of moss-microbial interactions, including aspects of community structure, taxon functional relationships, and finally more broadly ecology including the biogeochemical C-cycle, with the hope of encouraging more synoptic field-based and laboratory experimental studies of the synecology of mosses and their associated microbial communities. For conciseness in the text, shell-bearing, testate amoebae will be referred to as testacea. Non-shell-bearing, naked amoebae (previously known as gymnamoebae) will be referred to as amoebae, and heterotrophic nanoflagellates will be concisely referred to as nanoflagellates.

Microbial Taxonomy and Moss-Microbial Community Structure

Remarkably little research has been published in the primary literature on bryophyte-moss and microbial structural associations across broad categories of ecosystems. Structural associations are considered here to be analyses of the taxonomy, diversity and spatial organization of microbes inhabiting mosses. A rather thorough search of published work, including heterotrophic bacterial and protist taxa, using *Biosis* (*Biological Abstracts*) from 1926 to present, yielded only 42 relevant publications, and many of these included largely documentation of specific microbe taxa (e.g., testacea) or heterotrophic bacteria (e.g., 21 during 1969 -2013) that were associated with moss collected from particular sites. There was less published data on the diversity of microbes associated with different mosses. However, a much greater number of publications documented relationships of mosses with cyanobacteria including functional relationships and ecology. Some relevant examples will be included in subsequent sections.

A useful review of protozoan diversity in mosses has been published by Glime (2013), including general taxonomic descriptions, abundance and habitat specificity for a broad range of protozoan protists (amoebae, ciliates and flagellates). Historically, testacea have received more attention than other protists, probably because they are readily identified in freshly collected and preserved samples by their characteristic species-specific tests, and in general appear to be particularly abundant in terrestrial mosses. Taxonomic descriptions and abundances have been recorded from diverse geographic locales, including Antarctica where mosses are a major part of the otherwise sparse vegetation (e.g., Smith, 1974; Smith and Tearle, 1985). Numbers varied from less than 300 to approximately 1,200 g^{-1} fresh weight of moss. Other island geographic locales where testacea have been enumerated or identified include Anabon Island (Gulf of Guinea) (Del Pilar, 1963), Alaskan tundra (Anderson, 2008), temperate forest, Northeastern U. S. A. (see Figure 1), Balearic Islands (Del Pilar, 1965a), Canary Islands (Del Pilar, 1965b), South Georgia sub-Antarctica (Vincke et al., 2006), North-East Greenland (Van Kerckvoorde et al., 2000), and Devon Island Canadian Arctic (Beyens and Chardez, 1994). Additional observations of testacea at continental locations include Bulgaria (Davidova, 2008; Golamanski, 1964), Chile (Zapata, 2006), China (Bartos, 1963), Germany, Volga River basin (Mazei et al., 2007), Hungary (Ertl, 1960), Italian southeastern Alps (Mitchell et al., 2004), Karpathean Mountains (Bartos, 1940), Mexico (Sigala-Regalado et al., 2011), Norway (De Cloitre, 1962), and from dried moss collected at a variety of geographic locations during 80 years (Stepanek, 1963).

Structural studies of the association of testacea with moss are limited, but some examples are cited. Schönborn (1963) examined the stratigraphy of living testacea in sphagnum of peat bogs and identified three distinct layers:

the upper loose green layer contains chlorophyll-bearing testacea, the middle layer contains non-chlorophyllous species, and the third lowest layer (about 20 cm down) contained the empty shells of dead testacea of all kinds. The major species found in the uppermost layer were *Hyalosphenia papilio, Placocista spinosa, Heliopera sphagni, Amphitrema flavium and Amphitrema stenostoma.* The middle layer, largely decompositional, provided organic residues that supported a heterotrophic food web most typically characterized by the testacean *Hyalosphenia elegans*. Six distinct groups of testacean associations were found at increasing soil depths.

Fewer citations to published literature were found for association of other protists with moss. However, some examples are summarized. The densities of amoebae among *Hylocomium*-rich soil samples from Arctic tundra ranged from 1 to 3 x 10^4 g^{-1} dry weight (Anderson, 2008). The densities of the amoebae in samples from a Northeastern U.S.A. site (Anderson, 2006a) were slightly less, but comparable to that at the tundra sites. The mean number of different morphospecies at the northeastern site was 17. The number of heterotrophic nanoflagellates in the tundra moss-rich soil varied from 1 to 4 x 10^7 g^{-1} moss dry weight. Additional studies of the presence of flagellates in moss communities are limited, and examples include 19 species isolated from moss habitats in the Matveev and Dolgii Islands of the Pechora Sea (Tikhonenkov and Mazei, 2006/7), and 323 species or varieties tabulated from Manchuria, eastern Asia (Skvortzow, 1957). Tikhonenkov et al. (2012) identified over 40 species of heterotrophic flagellates associated with subtropical moss, litter and soil in Hubei Province of China in nine different habitats. Ciliates specifically have been examined in moss at some geographic locations (e.g., Bielanska-Grajner et al., 2011; Foissner, 1996a,b, 2000; Mieczan, 2009), but there appears to be less focus on moss-dwelling ciliates compared to other protists.

Heterotrophic bacteria associated with moss have received increasing attention in recent decades, especially with the application of modern molecular genetic and phospholipid analytic techniques. Analysis of phospholipids extracted from moss and sedge peat deposits showed significant differences in the composition of phospholipid fatty acids (PLFA) with respect to differences in vegetation. *Carex* peats in general had a higher proportion of Gram-negative bacteria, but lower proportion of Gram-positive bacteria compared to *Sphagnum* peats (Borga et al., 1994). A similar study (Sundh et al., 1997) at two boreal peatlands in Sweden, using principal-component analysis of PLFA, reported that there were substantial variations in PLFA composition among peatland habitats. There were general differences between wet sites and dry sites, but variation during the growing season was negligible. Principal-component analyses also showed that most PLFAs were in clusters of covarying fatty acids, suggesting that they originated in the same functional groups of microorganisms. Major clusters were formed by the following groups of PLFAs: monounsaturated (gram-negative eubacteria), terminally branched (gram-positive or anaerobic gram-negative eubacteria), methyl-branched and branched unsaturated (sulfate-reducing bacteria and/or actinomycetes), omega-8 monounsaturated (methane-oxidizing bacteria), and polyunsaturated (typical of eukaryotes). Moreover, the depth distributions in the peat differed. For example, PLFAs from sulfate-reducing bacteria and/or actinomycetes, and those from methane-oxidizing bacteria, had maximum concentrations either slightly below or near to the average water table depth, respectively.

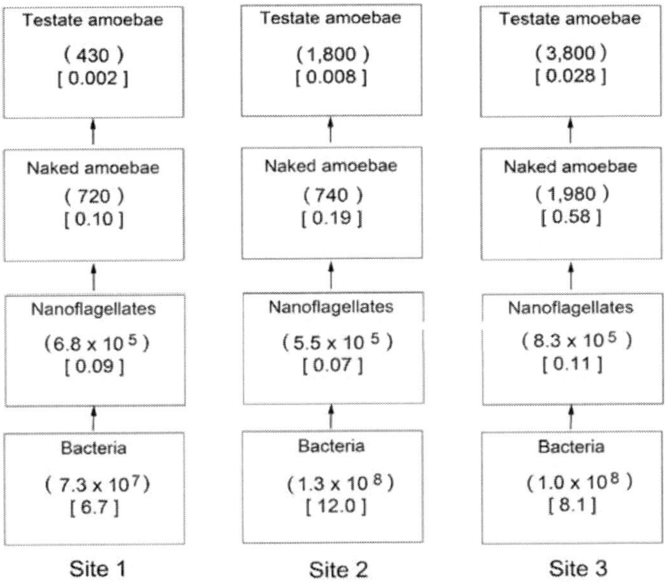

Figure 1. Comparative data for densities, number g^{-1} soil dry weight, (in parentheses), and C-biomass, μg g^{-1} soil dry weight, (in brackets) for bacteria, nanoflagellates, naked amoebae and testate amoebae ordered at ascending levels of the food chain for three sampling sites in deciduous tree stands. Site 1, moss on a rock in the forest floor under the tree canopy. Site 2, moss on the base of a tree at an illuminated northern edge of the tree stand. Site 3, a heavy moss mat on a fallen log at a well-illuminated, eastern edge of a tree stand.

In addition to environmental distribution patterns, the spatial distribution of bacteria in the upper green part and the lower brown part of the Antarctic moss *Sanionia uncinata* was examined using 16S rRNA gene pyrosequencing (Park et al., 2013). Proteobacteria were the most dominant (65.6%), followed by Bacteroidetes (29.1%), and Actinobacteria (11.7%). The difference in distribution of Alphaproteobacteria in the upper part (2%) and lower part (22.2%) of the moss was statistically significant. Moreover, dominant and diverse species were detected and closely related to the environmental sequences. Similarly, Hornschuh et al. (2002) examined the distribution of epiphytic bacteria on the leaf surfaces of *Funaria hygrometrica*. Bacteria on the phylloid surfaces were numerous in the grooves between adjacent lamina cells. Two methylobacterium strains produced a moss response similar to the well-established effect of laboratory application of cytokinins on bud formation in *Funaria* protonemata. Moreover, Methylbacterium inoculation promoted the growth of protonemal filaments, suggesting that bacterial-moss interactions may be important in moss adaptive capacity in varying environments.

Bacteria associated with moss rhizosphere in Bangladesh were isolated and analyzed by Khan et al. (1997), especially their effects on protonemal development. Fifteen bacterial and eight actinomycetes were studied from an initial isolation containing 84 bacterial and 24 actinomycete strains including *Bacillus*, *Sporolactobacillus*, *Micrococcus*, *Paracoccus*, *Frankia*, *Intrasporangium*, *Micromonospora*, *Promicromonospora*, and *Streptomyces*. Experiments using extracts of rhizoids showed positive stimulation of some bacterial isolates. In a complementary study, Speiss et al. (1981) examined the effects of bacteria associated with mosses on the growth and development of moss protonema and concluded that bacteria adhering to moss in nature may affect protonemal growth and gametophore initiation, and therefore may be a significant ecological factor in regulating normal life cycle events at early stages of moss development. In further laboratory studies, Speiss et al. (1986) isolated bacteria from three different moss species and examined their effects on protonemal growth and development when inoculated on other moss species. Among 283 bacterial isolates, 48% - 68% promoted development of their host moss protonema. However, samples of different mosses collected on the ground within a few meters of the host moss responded differently to these bacteria, possibly indicating a host-specific bacterial relationship. This was further supported by evidence that moss host plants of a particular species that were collected ca. 170 miles apart, showed the same host-specific bacterial enhancement effect, thus suggesting that the effect is species specific across a wide geographic range. Overall, however, their data indicated that habitat as well as the species of moss, and possibly major changes in locale, may influence the kinds of bacteria found adhering to the same particular moss in nature, as estimated by their ability to alter growth and development of the moss species.

Functional Relationships

There is evidence that moss-based microbial communities are organized in a trophic hierarchy similar to other biological communities, i.e. with producers (in this case bacteria) at the base and larger consumers ordered with increasing size and decreasing densities and/or C-biomass in higher levels of the hierarchy (e.g., Figure 1). These data were obtained for three moss samples collected from a deciduous forest site at Torrey Cliff, Palisades, N.Y. Sample 1 was from a moss-covered rock on the forest floor in the shaded understory of the tree canopy. Sample 2 was collected at the base of a tree exposed to more intense daylight at the northeastern perimeter of the tree stand. Sample 3 was obtained from a dense mat of moss growing on a log located at a well-illuminated eastern edge of a deciduous tree stand. In general, bacterial productivity increased across the three samples, probably due to the increased light exposure among the three sites, thus providing increasingly greater photosynthetic carbon sources to support the microbial community. A similar hierarchical trophic organization of microbes was found at an Alaskan tundra site (Anderson and McGuire, in press). The C-biomass of the microbes expressed as $\mu g\ g^{-1}$ soil dry weight decreased with increasing trophic level, namely bacteria (20 – 120), nanoflagellates (2 – 14), amoebae (0.13 – 0.94), and testacea (~ 0.02).

Moss plants provide a substantial source of organic nutrients to support a robust microbial community. For example, Ying-Di et al. (1999) reported that among three species of *Plagiomnium*, growing in western Hunan, China, photosynthesis rates were about 40 – 63 $\mu mol\ CO_2\ kg\ dw^{-1}\ sec^{-1}$, with a maximum at temperatures between 25 – 30 °C. Goulden and Crill (1997), studying the ecology of a black spruce forest, reported that the photosynthesis rates of feather moss (*Hylocomium splendens*) and sphagnum, expressed as concentration of photosynthate products, were 0.5 to 1.0 $\mu mol\ m^{-2}\ sec^{-1}$ and 0.5 to 2.5 $\mu mol\ m^{-2}\ sec^{-1}$, respectively. Likewise, Sommerkorn et al. (1999) reported that the photosynthesis of a moss layer in a wet tundra of Taimry Peninsula, Siberia, assimilated as much as 51 to 98% of the daily amount of

CO_2 released from the wetland soils. Soluble photosynthetic organic compounds leached from feather moss in subalpine spruce forests during precipitation events can be as large as 1,544 mg m^{-2} (Wilson and Coxson, 1999).

Bacteria and fungi are particularly efficient in mobilizing large molecular weight organic compounds, largely through decomposition processes (e.g., Adl, 2003). Bacteria rapidly assimilate available soluble organic carbon compounds (SOC), and are usually among the first biota in microbial communities to metabolize available carbon nutrients. Exoenzymes released by the bacteria mobilize smaller organic molecules from particulate organic carbon (POC) through extracellular digestion. The rapid growth and small size (micron range) of bacteria make them a major primary food source for phagotrophic microbial predators higher up the food web, including nanoflagellates, amoebae and testacea. Nanoflagellates in the 2 to 5 μm range are particularly effective predators on bacteria, followed in decreasing order by amoebae and testacea. Amoebae prey on nanoflagellates, smaller amoebae, and bacteria. Some vampyrellid amoebae also prey on fungi. The prey of testacea is less well documented. Typically they also consume bacteria, but given their large size range (tens to hundreds of microns), they most likely also consume smaller protists and possibly small invertebrates such as round worms.

As mentioned previously, a substantial interest has developed in the functional relationships among cyanobacteria and mosses. Much of the attention has been directed to the nitrogen-fixing role of cyanobacteria and contribution of nitrogen nutrients to their associated moss plants and the terrestrial environments in a variety of geographic locales including Antarctica (Pandey et al., 1992), Arctic (Solheim et al., 2004), boreal forests (DeLuca et al., 2002; Zackrisson et al., 2004), alpine tundra in Colorado (Wojciechowski and Heimbrook, 1984), northern Sweden forests (Gentili et al., 2005), 30 lake islands in Sweden (Ininbergs et al., 2011), Surtsey, Iceland (Rodgers and Henriksson, 1976), forests of Canada (Houle et al., 2006) and temperate U.S.A., including New Hampshire and North Carolina (Jean et al., 2012; Lambert and Reiners, 1979). Overall, major attention has been given to higher latitude locations, and much more research is warranted on cyanobacterial-moss associations in other moss-rich locales, including montane cloud forests, temperate and tropical rain forests, as well as a broader biogeographical analysis of cyanobacterial-moss associations in major ecosystems across high and low latitudes in both the northern and southern hemispheres.

Attention has been given to the dynamics of association between cyanobacteria and mosses in relation to abiotic and biotic factors. Some representative studies are reviewed. The rate of N-2 fixation by moss cyanobacteria has been measured among varying species and in a variety of environments. In boreal forests, seasonal N-2 fixation rates were estimated to be 193 mg N m^{-2} for *Sphagnum capillifolium* and 23 mg m^{-2} for *Pleurozium schreberi* (Markham, 2009). Zackrisson et al. (2009) studied the season-long assessment of N-2 fixation rates at a mixed feather moss site in northern Sweden and showed that *Hylocomium splendens* fixed a substantial quantity of N-2, but about 50% less total nitrogen compared to the contribution from *Pleurozium schreberi.* All-totaled, both species provided 1.6 kg fixed N-2 ha^{-1} $year^{-1}$. However, *H. splendens* had somewhat higher N-fixation rates at high fertility sites compared to *P. schreberi. Nostoc* spp. and *Stigonema* spp. were the primary cyanobacteria colonizing *H. splendens* and *P. schreberi,* as also reported by Gentili et al. (2005), who additionally found the cyanobacterium *Calothrix* sp. They also examined the influence of temperature as well as light and temperature together on N-2 fixation. The *Nostoc* sp. was found to fix N-2 at lower temperatures than the *Calothrix* sp.

The kinetics of short-term N-2 fixation by moss-associated cyanobacteria were studied by Jean et al. (2012) in warm (North Carolina) and cold (Canadian) temperate forests. Samples of the moss Anomodon attenuatus were collected at both sites during summer. They found that N-2 fixation by the moss carpet was strongly influenced by temperature, humidity, and light. N-2 fixation was highly dependent on temperature and showed rapid response in samples from both regions, reaching a maximum at a higher temperature for the North Carolina samples than those from Canada. While N-2 fixation in mosses from North Carolina was maintained up to 36° C, it was totally suppressed at 36° C in mosses from Canada. After cold treatment, N-2 fixation resumed in the samples from both regions within minutes with no observable hysteresis between 5 and 24° C. A similar decrease in N-2 fixation with increasing temperature in cyanobacterial-moss associations at a subantarctic island was reported by Smith (1984). He found that N-2 fixation increased from -5° C to a maximum at 25 – 27° C, but declined sharply at temperatures above the optimum and were lower at 35° C than at 0° C. Q10 values in a temperature range of 0 – 25° C were between 2.3 – 2.9, depending on photosynthetic photon flux density.

Jackson et al. (2011) studied the effects of precipitation patterns on N-2 fixation, and found that nitrogen fixation by cyanobacteria in feather mosses (the main biological nitrogen input to boreal forests) was strongly influenced by both moisture amount and frequency, and their interaction. Increased precipitation frequency had greater effects when amounts were higher. Within a given amount of moisture, nitrogen fixation varied up to seven-fold depending on temporal distribution. Overall, their results highlighted that both the amount and temporal distribution of rainfall determine the effect of feather mosses on ecosystem N input and the decomposer subsystem. Similar results of the positive effects of available water on N-2 fixation in the Norwegian high Arctic were reported by Zielke et al. (2005). At six sites varying in types of moss-dominated vegetation, nitrogen fixation activity was positively correlated with the amount of available water in the vegetation, At two sites, where water content of the vegetation was constantly higher than 80% (w/w) throughout the season, N-2 fixation activity was correlated with temperature.

DeLuca et al. (2007) studied the changes in cyanobacterial N-2 fixation in boreal forest moss stands in relation to ecological successional stages using a reciprocal transplant experiment in early and late successional forests. They found that moss transplanted to late successional forests exhibited high rates of N-2 fixation and had consistently high numbers of cyanobacteria on moss shoots, but low levels of available nitrogen. Conversely, transplants in early successional forests had higher nitrogen availability but low rates of N-2 fixation, and there was limited presence of cyanobacteria on moss shoots. Zackrisson et al. (2004) also reported that N-2 fixation in boreal forests becomes more important in late succession. They concluded that limited nitrogen availability in late succession is clearly one of the primary drivers of N-2 fixation rates in boreal forest ecosystems, and they proposed that these findings might help to explain the origin of high rates of net nitrogen accumulation in soil unaccounted for previously at northern boreal sites.

Evidence of moss-associated cyanobacterial fixation of N-2 raises the question of how much of the fixed nitrogen is utilized by the moss and to what extent it may support moss growth. Berg et al. (2013) examined the fate of fixed nitrogen in *Sphagnum riparium* in laboratory cultures for 57 days, with and without cyanobacteria. They found that nitrogen fixation by cyanobacteria associated with *Sphagnum* can promote moss growth by transfer of fixed nitrogen to the moss. More than 35 % of the nitrogen fixed by the cyanobacteria was transferred to the newly formed moss biomass resulting in an increase in the growth of *Sphagnum* biomass compared to the controls. The variation in the increase of nitrogen content explained 76 % of the biomass increment. Therefore, nitrogen fixation can have an immediate effect on the carbon fixation by *Sphagnum*, and suggests cyanobacterial symbioses with mosses may have a significant role in coupling of the nitrogen and carbon cycles in moss communities.

Moss-Microbial Community Ecology and the Biogeochemical Carbon Cycle

Given the current published research on functional relationships among moss species and various microbial taxa, there appears to be potential for significant synoptic ecological research on moss microbial communities. However, there are few published papers on the synecology of microbial communities associated with mosses. This is in marked contrast to the more substantial research on general moss ecology and global biogeography (e.g., Casanovas et al., 2013; Geffert, et al., 2013; Glime, 2007; O'Neill, 2000). Most recent research on moss microbial ecology is at high latitudes, especially Arctic tundra, where the effects of climate change and global warming are expected to have major effects, especially in relation to moss primary production, microbial respiratory CO_2 release and the biogeochemical C-cycle. Information on the major issues related to high latitude environments and climate change can be found in recent reviews (e.g., Gutierrez and Pena, 2010; Oechel et al., 1993).

A significant amount of organic C is stored in the permafrost as moss-derived peat deposits in the tundra. With increased global warming, and permafrost thawing, these organic compounds, in addition to photosynthates eluted from the living moss, become available for microbial metabolism and respiratory CO_2 production. Ecological research has been published on one or more particular microbial taxa in relation to high latitude ecosystems, including bacteria (e.g., Mänistö et al., 2013), fungi (e.g., Davey, et al., 2013; Osono et al., 2012), selected groups of protists (e.g., Tsyganov, 2013), and other examples reported in previous sections of this chapter. Some attention has been given to integrating analyses of tundra bacteria and fungi in relation to varying habitats and climate change (e.g., Scmidt and Bölter, 2002; Rinnan and Bååth, 2009; Wallenstein et al., 2007).

This portion of the review is focused on the ecology and C-biogeochemistry of bacteria, fungi and protists taken collectively in moss-rich, high-latitude environments, especially recent contributions. Sistla et al. (2013) reported the results of a two-decade summer warming experiment in an Alaskan tundra ecosystem. Data comparing artificially warmed plots to controls showed that warming increased plant biomass and woody dominance, indirectly increased winter soil temperature, homogenized the soil trophic structure across soil horizons and suppressed surface-soil-decomposer activity. However, the warming did not change total soil carbon or nitrogen stocks, thereby increasing net ecosystem belowground carbon storage. Trophic structure across the moss-rich soil horizons was partially determined by canonical discriminant analysis of protists (amoebae, flagellates and ciliates grouped together), indicating that the protists in the warming treatment were more homogeneously distributed across soil horizons compared to the control treatment. The surface organic soil carbon and nitrogen stocks were unaffected by warming, although the C:N ratio declined, reflecting a decline in carbon-rich mosses and lichens and greater accumulation of leaf litter. In the deep organic horizon of the treatment plot (5 cm below the surface to mineral soil horizon), both bacterivores and the ratio of fungal to bacterial biomass declined significantly, indicating that warming restricted both the root- and detritus-fed trophic pathways. No attempt was made to account for the C-content of the individual protist groups, but bacteria and fungi accounted for most of the C biomass.

The effects of nitrogen amendment on bacteria and protists in tundra soil were studied by Stapleton et al. (2005). Analyses over three summers showed that although bacterial biomass did not change substantially with N enrichment, bacterial production (measured by ^{3}H thymidine incorporation into DNA) increased by the second summer in NH_4 amended plots, but not in NO_3 amended plots. The main bacterial predators were heterotrophic flagellates and naked amoebae, but they showed no significant response to the N addition. The flagellates showed a correlation with bacterial abundance, suggesting a dependence on bacteria as a food source, a finding corroborated in studies of moss-rich tundra soil samples (Anderson, 2013), where the correlation of bacteria and heterotrophic nanoflagellates obtained for 14 transect samples was positive and statistically significant ($r = 0.75$).

With increasing evidence of climate change and global warming due to atmospheric CO_2 concentrations, interest has arisen in the role of terrestrial microbial communities in the carbon cycle (e.g., Anderson 2010a), especially as a carbon source (respiratory CO_2) and sink (e.g., consumer of organic carbon nutrients). Some recent research in the Arctic tundra is reviewed. Fungi and bacteria are the most abundant microbes and account for the most organic C among microbial communities in some Arctic tundra sites. For example in tundra moss-rich soil samples collected in early to mid summer, fungal C-biomass ranged from 5 to 11 mg g^{-1} soil dry weight, and bacteria accounted for 23 to 58 $\mu g\ g^{-1}$ soil dry weight (Anderson and McGuire, in press). By contrast the C-biomass among the protists, expressed as μg g^{-1} was much smaller, namely, heterotrophic flagellates (2 to 7), amoebae (0.13 to 0.40), and testacea (0.002 to 0.012), exhibiting a decreasing hierarchical relationship similar to that reported for low latitude moss-rich associations (e.g., Figure 1). When the samples were enriched with soluble organic carbon (SOC) in the form of a 1% (w/v) glucose solution, fungal C-biomass showed no significant increase after seven days of treatment, possibly due to the small sporadic infusion of SOC and/or insufficient time to affect fungal nutrition. However, bacteria and protist C-biomass increased substantially, i.e. bacteria (122 – 214%), heterotrophic flagellates (200 – 230%), amoebae (200 – 231%) and testacea (143% to tenfold). Prior research at other Arctic tundra sites has also confirmed the same general pattern of C-biomass for bacteria and protists in freshly collected moss-rich soil samples near Toolik Lake, AK (Anderson, 2008, 2010a). However, these samples collected in late summer gave evidence of much higher C-biomass estimates (μg g^{-1}) for each microbial taxon, namely bacteria (80 – 400), heterotrophic flagellates (13 – 53), amoebae (0.6 – 3.0) and testacea (6 – 40), the latter likely due to the presence of larger species compared to those sampled in the study done in early to mid-summer. However, as is typical for trophic hierarchies, the densities of biota (number g^{-1}) decreased with increasing levels, bacteria (10^9 – 10^{10}), heterotrophic flagellates ($1 – 4 \times 10^7$), amoebae ($2 – 3 \times 10^4$) and testaea ($1 – 6 \times 10^3$).

Estimation of the amount of C consumed by microscopic predators in microbial food webs is challenging, partially due to the large range in size (several μm to tens of μm) and lack of information on the range of prey consumed, especially for larger predators in upper trophic levels. However, many of the microbial predators consume bacteria as their primary source of nutrition, or they prey on smaller protists that most certainly are obligate bacterial predators due to their small size (e.g., heterotrophic nanoflagellates). This makes the food chain somewhat less complex than those with larger predators such as invertebrates. The consumption rate of bacteria within moderate temperature ranges of ~ 25° C has been estimated fairly accurately for heterotrophic nanoflagellates to be between 10 and 20 bacteria $flagellate^{-1}$ h^{-1}. For amoeba, bacterial predation rates are estimated using size-based regression equations (e.g., Anderson, 2006b; Anderson and McGuire, in press; Rogerson et al., 1996). For example, amoebae in the moderate size range of 20 to 30 μm are estimated to consume between 100 and 220 bacteria $amoeba^{-1}$ h^{-1}. The grazing pressure of small heterotrophic flagellates on bacteria is greater than other protists that are at higher levels of the microbial food chain because of the sheer abundance and obligate bacterivory of the nanoflagellates. These estimations are only approximate, largely due to the narrow temperature range for the consumption data used to derive the equations and also in some cases the limited number of species used to establish the mathematical relationships, although the size ranges used may be sufficiently broad.

One of the important considerations regarding the role of tundra microbial communities in the C-cycle is to account for how much of the C that is consumed by the microbes is eventually incorporated into particulate biomass as opposed to release as respiratory CO_2. Field data on rates of microbial production of CO_2 are limited. Some recent laboratory experimental data are reviewed here. Alaskan tundra soil samples were experimentally amended with glucose solution, or with an equivalent amount of pure water in controls, to determine how much of the added soluble C was incorporated into microbial biomass during seven days of culture and how much was released as CO_2 (Anderson, 2012). Two samples of moss-rich tundra soil were taken in spring and one in summer for the experiment. Addition of 1% (w/v) glucose solution produced an immediate, pronounced two- to three-fold increase in respiration above basal rate, which declined over 4 hours to baseline levels. Less than 1% (w/w) of glucose-C supplement was respired during the respiratory spike, relative to the 89 $\mu g\ g^{-1}$ added. A more substantial amount of the glucose-C became incorporated in microbial biomass. The total difference in microbial carbon ($\mu g\ g^{-1}$) between the experimental treatments and controls without glucose after 1 week, for the two spring samples and one summer sample, was as follows: spring sample 1 (8), spring sample 2 (31), and summer sample (70). The percent of glucose-C incorporated for each sample was spring sample 1 (5%), spring sample 2 (17%), and summer sample (39%). Most of it was attributed to biomass of bacteria and heterotrophic nanoflagellates. Although the respiratory response to pulsed glucose-C was minimal as reported above, the overall mean basal respiratory rate after 1 week ranged between 4 to 6 nmol $min^{-1}\ g^{-1}$ soil, indicating a significant assimilation and respiration of constituent soil organic carbon. Eventually, some of the biomass C is incorporated into organically complex and relatively stable biological material during subsequent passage through the food web, including non-living products such as cyst walls, chitinous exoskeletons, and other relatively stable organic residues. Thus, during transfer up the food chain, a considerable amount of microbial assimilated C may eventually be sequestered in relatively stable biological indurated matter, in addition to some loss by respiratory CO_2 emissions by predatory biota at higher trophic levels.

Terrestrial emissions of CO_2, largely contributed by biological respiration from plant roots and terrestrial microbes, have been assessed by atmospheric CO_2 sensors and monitoring devices of various kinds (e.g., Vourlitis and Oechel, 1999; Vourlitis et al., 2000). However, there are few convenient means for estimating total microbial respiratory CO_2 emissions. Given the large densities of bacteria in moss-rich soils, their substantial role in SOC metabolism, and the strong correlation of bacterial numbers with fungal and nanoflagellate biomass, it is very likely that counts of bacteria in small soil samples may be a good proxy to estimate microbial respiratory CO_2 production. Recently, data on CO_2 emissions from moss-rich tundra soil sampled along a 10 km transect in northwestern Alaska have shown that bacterial counts in thoroughly mixed small soil samples, with root material removed, are a good predictor of total microbial CO_2 production based on laboratory measurements (Anderson, 2013). A linear relationship between soil respiration and bacterial densities was obtained for two temperatures (10° and 20° C). The regression equation at 10° C is $R_s = 10.25 + 6.26\ B_d$, and at 20° C is $R_s = 12.32 + 14.07\ B_d$; where R_s is soil respiration (nmol $min^{-1}\ g^{-1}$ soil dry weight) and B_d = bacterial density (number of bacteria x $10^9\ g^{-1}$ soil dry weight). The mean ± s.e. for the Q_{10} of the respiration rate was 2.04 ± 0.10, a value comparable to the mean (2.05) reported for two other Arctic tundra sites (Anderson 2010b). Densities of bacteria varied from 2.7 – 16 x 10^9, and flagellates 0.7 – 7.9 x 10^7. C-biomass (µg g^{-1} soil) of bacteria varied from 358 to 2,114, and flagellates 12 – 37.

Conclusion

Moss communities at high and low latitudes provide a rich source of nutrients supporting a diverse microbial community including photosynthetic, nitrogen-fixing, cyanobacteria and heterotrophic bacteria at the base of a hierarchically organized microbial food web with larger heterotrophic protozoa at the highest levels. The larger protozoa prey upon bacteria and smaller protists lower in the trophic hierarchy. Cyanobacterial-fixed N can be transferred to the moss host as nutrients, thus enhancing moss growth. In addition to nutrient richness derived from moss primary production, the amount of precipitation and available moisture is an important environmental forcing function that, in conjunction with temperature, affects the abundance and productivity of moss-dwelling cyanobacteria and heterotrophic bacteria, thus influencing the complexity and total productivity of the microbial food webs. Laboratory experiments using pulses of low-molecular-weight organic nutrients, such as those secreted from mosses, show that pulsed addition of organic nutrients produces a brief burst of microbial respiration and release of respiratory CO_2 into the atmosphere. However, as much as 99% of the C in the nutrients may be sequestered within the microbial particulate biomass, or bound in the soil particles, and may not immediately contribute significantly to release of climate-warming CO_2 to the atmosphere.

Presently, we have very limited information on microbial ecology and the role of microbial communities in the C-cycle at moss-rich sites including the tundra, where global warming is presently most pronounced, or other geographic locales, and much more broad geographical research is needed. Our knowledge of the complex way that mosses may serve as a biological buffer in the C-cycle, and mitigation of atmospheric CO_2 concentrations through photosynthetic uptake relative to biological respiration, is currently meager. This is also the case on a broader global scale. However, given the geographic vastness of moss-rich soils in high latitude boreal and tundra ecosystems, as well as in other terrestrial regimes, much more ecological research is warranted to more fully appreciate the C-biogeochemical role of moss-based biological communities on a global scale.

ACKNOWLEDGMENT

I thank Prof. Kevin Griffin for reviewing this contribution and making helpful recommendations. Some of the research by the author reviewed here was supported partially by funds from the United States National Science Foundation. This is Lamont-Doherty Earth Observatory Contribution Number 7746.

REFERENCES

Adl, S. M. (2003). *The Ecology of Soil Decomposition.* Oxford: CAB International Publishing.

Anderson, O. R. (2006a). The Density and Diversity of Gymnamoebae Associated with Terrestrial Moss Communities (Bryophyta: Bryopsida) in a Northeastern U.S. Forest. *Journal of Eukaryotic Microbiology, 53,* 275-279.

Anderson, O. R. (2006b). A method for estimating cell volume of amoebae based on measurements of cell length of motile forms: Physiological and ecological applications. *Journal of Eukaryotic Microbiology*, *53*, 185-187.

Anderson, O. R. (2008). The Role of amoeboid protists and the microbial community in moss-rich terrestrial ecosystems: Biogeochemical implications for the carbon budget and carbon cycle, especially at higher latitudes. *Journal of Eukaryotic Microbiology*, *55*, 145-150.

Anderson, O. R. (2010a). The reciprocal relationships between high latitude climate changes and the ecology of terrestrial microbiota: Emerging theories, models, and empirical evidence, especially related to global warming. In: B. Gutierrez & C. Pena (Eds.), *Tundras: Vegetation, Wildlife and Climate Trends* (47-79). New York: Nova Science Publishers.

Anderson, O. R. (2010b). An Analysis of Respiratory Activity, Q_{10}, and Microbial Community Composition of Soils from High and Low Tussock Sites at Toolik, Alaska. *Journal of Eukaryotic Microbiology*, *57*, 218-219.

Anderson, O. R. (2012). The Fate of Organic Sources of Carbon in Moss-rich Tundra Soil Microbial Communities: A Laboratory Experimental Study. *Journal of Eukaryotic Microbiology*, *59*, 564-570.

Anderson, O. R. (2013). Bacterial and heterotrophic nanoflagellate densities and C-biomass estimates along an Alaskan tundra transect with prediction of respiratory CO_2 efflux. *Journal of Eukaryotic Microbiology*, retrieved 15 Oct. 2013 at doi:10.1111/jeu.12081.

Anderson, O. R. & McGuire, K. (in press). C-biomass of bacteria, fungi, and protozoan communities in Arctic tundra soil, including some trophic relationships. *Acta Protozoologica*.

Bartos, E. (1940). Studien über die moos-bewohnenden Rizopoden der Karpathen. *Archiv fur Protistenkunde*, *94*, 93-160.

Bartos, E. (1963). The rhizopods from some moss samples from China. *Vestnik Ceskoslovenske spolecnosti zoologicke*, *27*, 85-96.

Berg, A., Danielsson, A. & Svensson, B. H. (2013). Transfer of fixed-N from N-2 fixing cyanobacteria associated with the moss *Sphagnum riparium* results in enhanced growth of the moss. *Plant & Soil*, *362*, 271-278.

Beyens, L. & Chardez, D. (1994). On the habitat specificity of the testate amoebae assemblages for Devon Island (NWT, Canadian Arctic), with description of a new species: *Difflugia ovalisina*. *Archiv fuer Protistenkunde*, *144*, 137-142.

Bielanska-Grajner, I., Mieczan, T. & Cudak, A. (2011). Co-occurrence of ciliates and rotifers in peat mosses. *Polish Journal of Environmental Studies*, *20*, 533-540.

Borga, P., Nilsson, M. & Tunlid, A. (1994). Bacterial communities in peat in relation to botanical composition as revealed by phospholipid fatty acid analysis. *Soil Biology & Biochemistry*, *26*, 841-848.

Bowker, M. A. (2007). Biological soil crust rehabilitation in theory and practice: An underexploited opportunity. *Restoration Ecology*, *15*, 13-23.

Casanovas, P., Lynch, H. J. & Fagan, W. F. (2013). Multi-scale patterns of moss and lichen richness on the Antarctic Peninsula. *Ecography*, *36*, 209-219.

Clements, F. E. (1916). *Plant succession: analysis of the development of vegetation*. Carnegie Institute, Washington, D. C.

Davey, M. L., Einar, H., Rune, H., Håvord, K. & Ohlson, M. (2013). Amplicon-pyrosequencing-based detection of fungal communities along an elevational gradient. *Molecular Ecology*, *22*, 368-383.

Davidova, R, D. (2008). A study of the moss testate amoebae (Protozoa: Testacea) of the Strandzha Natural Park (South-Eastern Bulgaria). *Acta Zoologica Bulgarica*, *60*, 23-30.

De Cloitre, L. (1962). On some thecamoebae collected in Norway. *Hydrobiologia, 19*, 179-182.

Del Pilar, G. M. (1963). Results of the Peris-Alvarez expedition to the Island of Annobon (Gulf of Guinea). VI. Moss dwelling thecamoebae. *Publicaciones del Instituto de Biologia Aplicada*, *34*, 5-16.

Del Pilar, G. M. (1965a). Moss dwelling thecamoebae from Ibiza (Balearic Islands). *Publicaciones del Instituto de Biologia Aplicada*, *39*, 109-111.

Del Pilar, G. M. (1965b). On moss dwelling thecamoebae from Tenereife. *Publicaciones del Instituto de Biologia Aplicada*, *39*, 123-127.

DeLuca, T. H., Zackrisson, O., Gentili, F., Sellstedt, A. & Nilsson, M-C. (2007). Ecosystem controls on nitrogen fixation in boreal feather moss communities. *Oecologia*, *152,* 121-130.

DeLuca, T. H., Zackrisson, O., Nilsson, M-C., & Sellstedt, A. (2002). Quantifying nitrogen-fixation in feather moss carpets of boreal forests. *Nature*, *419*, 917-920.

Ertl, M. (1960). Contributions of knowledge of the moss inhabiting Thecamoebae of Hungary. *Opuscula Zoologica*, *4*, 31-37.

Foissner, W. (1996a). Faunistics, taxonomy and ecology of moss and soil ciliates (Protozoa, Ciliophora) from Antarctica, with description of new species, including *Pleuroplitoides smithi* gen. n., sp. n. *Acta Protozoologica, 35*, 95-123.

Foissner, W. (1996b). Terrestrial ciliates (Protozoa, Ciliophora) from two islands (Gough, Marion) in the southern oceans, with description of two new species, *Arcuospathidium cooperi* and *Oxytricha ottowi*. *Biology & Fertility of Soils*, *23*, 282-291.

Foissner, W. (2000). A compilation of soil and moss ciliates (Protozoa, Ciliophora) from Germany, with new records and descriptions of new and insufficiently known species. *European Journal of Protistology*, *36*, 253-283.

Geffert, J. L., Frahm, J.-P., Barthlott, W. & Mutke, J. (2013). Global moss diversity: spatial patterns of species richness. *Journal of Bryology*, *35*, 1-11.

Gentili, F., Nilsson, M-C., Zackrisson, O., DeLuca, T. H. & Sellstedt, A. (2005). Physiological and molecular diversity of feather moss associative N-2-fixing cyanobacteria. *Journal of Experimental Botany*, *56*, 3121-3127.

Glime, J. M. 2007. *Bryophyte Ecology. Volume 1. Physiological Ecology*. Ebook sponsored by Michigan Technological University and the International Association of Bryologists. Accessed on 14 October, 2013 at http://www.bryoecol.mtu.edu.

Glime, J. M. (2013). *Bryophyte Ecology*. Chapter 2-1. E-book sponsored by Michigan Technological University and the International Association of Bryologists. Accessed on 14 October, 2013 at http://www. bryoecol.mtu.edu.

Golamanski, V. (1964). Study of species and ecology of rhizopods (Rhizopoda, Testacea) in the peat-moss of Vitocha Mountain. *Godisnik na Sofijskiâ Universitet "Kliment Ohridski" Biologiceski Fakultet.Kniga 1, Zoologiâ*, *59*,17-38.

Goulden, M. L. & Crill, P. M. (1997). Automated measurements of CO_2 exchange at the moss surface of a black spruce forest. *Tree Physiology*, *17*, 537-542.

Gutierrez, B. & Pena C. (Eds.). (2010). *Tundras: Vegetation, Wildlife and Climate Trends*. New York: Nova Science Publishers.

Hornschuh, M., Grotha, R. & Kutschera, U. (2002). Epiphytic bacteria associated with the bryophyte *Funaria hygrometrica*: Effects of *Methylobacterium* strains on protonema development. *Plant Biology*, *4*, 682-687.

Houle, D., Gauthier, S. B., Paquet, S., Planas, D. & Warren, A. (2006). Identification of two genera of N-2-fixing cyanobacteria growing on three feather moss species in boreal forests of Quebec, Canada. *Canadian Journal of Botany*, *84*,1025-1029.

Ininbergs, K., Bay, G., Rasmussen, U., Wardle, D. A. & Nilsson, M-C. (2011). Composition and diversity of nifH genes of nitrogen-fixing cyanobacteria associated with boreal forest feather mosses. *New Phytologist*, *192*, 507-517.

Jackson, B. G., Martin, P., Nilsson, M-C. & Wardle, D. A. (2011). Response of feather moss associated N-2 fixation and litter decomposition to variations in simulated rainfall intensity and frequency. *Oikos*, *120*, 570-581.

Jean, M-E., Cassar, N., Setzer, C. & Bellenger, J.-P. (2012). Short-term N-2 fixation kinetics in a moss-associated cyanobacteria. *Environmental Science & Technology*, *46*, 8667-8671.

Khan, M. R., Huq, S. M. I. & Hasanuzzaman, M. (1997). Moss rhizosphere and its microflora. *Bangladesh Journal of Botany*, *26*, 163-168.

Lambert, R. L. & Reiners, W. A. (1979). Nitrogen fixing moss associations in the subalpine zone of the white mountains New-Hampshire USA. *Arctic & Alpine Research*, *11*, 325-334.

Mänistö, M. K., Kurhela, E., Tirola, M. & Häggblom, M. M. (2013). Acidobacteria dominate the active bacterial communities of Arctic tundra with widely divergent winter-time snow accumulation and soil temperatures. *FEM Microbiology Ecology*, *84*, 47-59.

Markham, J. H. (2009). Variations in moss-associated nitrogen fixation in boreal forest stands. *Oecologia*, *161*, 353-359.

Mazei, Y. A., Tsyganov, A. N., & Bubnova, O. A. (2007). The species composition, distribution, and structure of a testate amoeba community from a moss bog in the middle Volga river basin. *Zoologicheskii Zhurnal*, *86*, 1155-1167.

Mieczan, T. (2009). Ciliates in *Sphagnum* peatlands: Vertical micro-distribution, and relationships of species assemblages with environmental parameters. *Zoological Studies*, *48*, 33-48.

Mitchell, E. A., Bragazza, L. & Gerdol, R. (2004). Testate amoebae (Protista) communities in Hylocomium splendens (Hedw.) B. S. G. (Bryophyta): Relationships with altitude, and moss elemental chemistry. *Protist*, *155*, 423-436.

Oechel, W., Hastings, S., Vourlitis, G. & Jenkins, M. (1993). Recent change of Arctic tundra ecosystems from a net carbon dioxide sink to a source. *Nature*, *361*, 520 – 523.

O'Neill, K. P. (2000). Role of bryophyte-dominated ecosystems in the global carbon budget. In: A. J. Shaw & B. Goffinet (Eds.), *Bryophyte Biology,* (344-368). Cambridge, UK, Cambridge University Press.

Osono, T., Ueno, T., Uchida, M. & Kanda, H. (2012). Abundances and diversity of fungi in relation to chemical changes in arctic moss profiles. *Polar Science*, *6*, 121-131.

Pandey, K. D., Kashyap, A. K. & Gupta, R. K. (1992). Nitrogen fixation by cyanobacteria associated with moss communities in Schirmacher Oasis, Antarctica. *Israel Journal of Botany*, *41*, 187-198.

Park, M., Lee, H., Hong, S. G. & Kim, O-S. (2013). Endophytic bacterial diversity of an Antarctic moss, *Sanionia uncinata*. *Antarctic Science*, *25*, 51-54.

Rayburn, W. R., Mack, R. N, & Metting, B. (1982). Conspicuous algal colonization of the ash from Mount St. Helens. *Journal of Phycology*, *18*, 537–543.

Rinnan, R. & Bååth, E. (2009). Differential utilization of carbon substrates by bacteria and fungi in tundra soil. *Applied and Environmental Microbiology*, *75*, 3611-3620.

Rodgers, G. A. & Henriksson E. (1976). Associations between the blue-green algae *Anabaena-variabilis* and *Nostoc-muscorum* and the moss *Funaria-hygrometrica* with reference to the colonization of Surtsey Iceland. *Acta Botanica Islandica*, *4*, 10-15.

Rogerson, A., Hannah, F. & Gothe, G. (1996). The grazing potential of some unusual marine benthic amoebae feeding on bacteria. *European Journal of Protistology*, *32*, 271-279.

Schmidt, N. & Bölter, M. (2002). Fungal and bacterial biomass in tundra soils along an arctic transect from Taimyr Peninsula, central Siberia. *Polar Biology*, *25*, 871-877.

Schönborn, W. (1963). Die Stratigraphie lebender Testaceen im Sphagnetum der Hochmoore. *Limnologica 1*, 315-321.

Sigala-Regalado, I., Mayén-Estrada, R. & Morales-Malacara, J. B. (2011). Spatial and temporal distribution of protozoa at Cueva de Los Riscos, Querétaro, México. *Journal of Cave and Karst Studies*, *73*, 55-62.

Sistla, S. A., Moore, J., C., Simpson, R. T., Gough, L., Shaver, G. R. & Schimel, J. P. (2013). Long-term warming restructures Arctic tundra without changing net soil carbon storage. *Nature*, *497*, 615-618.

Skvortzow, B. W. (1957). New and rare flagellatae from Manchuria, eastern Asia. *Philippine Journal of Science*, *86*, 139-202.

Smith, H. G. (1974). A comparative study of protozoa inhabiting *Drepanocladus* moss carpet in the South Orkney Islands. *Antarctic Survey Bulletin*, *38*, 1-16.

Smith, H. G. & Tearle, P. V. (1985). Aspects of microbial and protozoan abundances in Signy Island fellfields. *Antarctic Survey Bulletin*, *68*, 83-90.

Smith, V. R. (1984). Effects of abiotic factors on acetylene reduction by cyanobacteria epiphytic on moss at a subantarctic island. *Applied & Environmental Microbiology*, *48*, 594-600.

Solheim, B., Wiggen, H., Roberg, S. & Spaink, H. P. (2004). Associations between Arctic cyanobacteria and mosses. *Symbiosis, 37*, 169-187.

Sommerkorn, M., Boelter, M. & Kappen, L. (1999). Carbon dioxide fluxes of soils and mosses in wet tundra of Taimyr Peninsula, Siberia: Controlling factors and contribution to net system fluxes. *Polar Research*, *18*, 253-260.

Speiss, L. D., Lippincott, B. B. & Lippincott, J. A. (1981). Bacteria isolated from moss and their effect on moss development. *Botanical Gazette*, *142*, 512-518.

Speiss, L. D., Lippincott, B. B. & Lippincott, J. A. (1986). Specificity of moss response to moss-associated bacteria, some influences of moss species habitat and locale. *Botanical Gazette*, *147*, 418-424.

Stapleton, L. M., Crout, N. M. J., Säwström, C., Marshall, W. A., Poulton, P. R., Tye, A. M. & Laybourn-Parry, J. (2005). Microbial carbon dynamics in nitrogen amended Arctic tundra soil: Measurement and model testing. *Soil Biology & Biochemistry*, *37*, 2088-2098.

Stepanek, M. (1963). Rhizopoda from old, dried-out moss samples. *Hydrobiologia*, *21*, 304-327.

Sundh, I., Nilsson, M. & Borga, P. (1997). Variation in microbial community structure in two boreal peatlands as determined by analysis of phospholipid fatty acid profiles. *Applied & environmental Microbiology*, *63*, 1476-1482.

Taylor, A. M. (1920). Ecological Succession of Mosses. *Botanical Gazette*, *69*, 449-491.

Tikhonenkov, D. V. & Mazei, Y. A. (2006/7). Heterotrophic flagellates from freshwater biotopes of Matveev and Dolgii Islands (the Pechora Sea). *Protistology*, *4*, 327-337.

Tikhonenkov, D. V., Mylnikov, A. P., Gong, Y. C., Feng, W. S. & Mazei, Y. (2012). Heterotrophic flagellates from freshwater and soil habitats in subtropical China (Wuban Area, Hubei Province). *Acta Protozoologica*, *51*, 65-79.

Tsyganov, A. N., Milbau, A. & Beyens, L. (2013). Environmental factors influencing soil testate amoebae in herbaceous and shrubby vegetation along an altitudinal gradient in subarctic tundra (Abisko, Sweden). *European Journal of Protistology*, *49*, 238-248.

Van Kerckvoorde, A., Trappeniers, K., Chardez, D., Nijs, I. & Beyens, L. (2000). Testate amoebae communities from terrestrial moss habitats in the Zackenberg area (North-East Greenland). *Acta Protozoologica*, *39*, 27-33.

Vincke, S., Van de Vijver, B., Gremmen, N. & Eyens, L. (2006). The moss dwelling testacean fauna of the Stromness Bay (South Georgia). *Acta Protozoologica*, *45*, 65-75.

Vourlitis, G. L. & Oechel, W. C. (1999). Eddy covariance measurements of CO_2 and energy fluxes of an Alaskan tussock tundra ecosystem. *Ecology*, *80*, 686–701.

Vourlitis, G. L., Harazono, Y., Oechel, W. C., Yoshimoto, M. & Mano, M. (2000). Spatial and temporal variations in hectare-scale net CO_2 flux, respiration and gross primary production of Arctic tundra ecosystems. *Functional Ecology*, *14*, 203–214.

Wallenstein, M. D., McMahon, S. & Schimel, J. (2007). Bacterial and fungal community structure in Arctic tundra tussock and shrub soils. *FEMS Microbial Ecology*, *59*, 428-435.

Walker, D. A. & Walker, M. D. (1996). Terrain and vegetation of the Imnavait Creek watershed. In: J. F. Reynolds, & J. D. Tenhunen (Eds.), *Landscape Function and Disturbance in Arctic Tundra* (73-108). Berlin: Springer.

Wilson, J. A. & Coxson, D. S. (1999). Carbon flux in a subalpine spruce-fir forest: pulse release from Hylocomium splendens feather-moss mats. *Canadian Journal of Botany*, *77*, 564-569.

Wojciechowski, M. F. & Heimbrook M. E. (1984). Di-nitrogen fixation in alpine tundra, Niwot Ridge Front Range Colorado, USA. *Arctic & Alpine Research*, *16*, 1-10.

Wynn-Williams, D. D. (1993). *Microbial processes and initial stabilization of fellfield soil.* In: J. Miles, & D. W. H. Walton (Eds.), *Primary Succession on Land* (17–32). Oxford, UK: Blackwell Scientific Publishers.

Ying-Di, L., Jing, L., Jinjun H. & Tong C. (1999). The measurement of net photosynthesis of three species of *Plagiomnium* mosses and its relationship to the light and temperature. *Journal of the Hattori Botanical Laboratory*, *87*, 315-324.

Zackrisson, O., DeLuca, T. H., Gentili, F., Sellstedt, A. & Jäderlund, A. (2009). Nitrogen fixation in mixed *Hylocomium splendens* moss communities. *Oecologia*, *160*, 309-319.

Zackrisson, O., DeLuca, T. H., Nilsson, M.-C., Sellstedt, A. & Berglund, L. M. (2004). Nitrogen fixation increases with successional age in boreal forests. *Ecology*, *85*, 3327-3334.

Zapata M. J. (2006). Thecamoebians (Protozoa, Rhizopoda) from the Rucapihuel peatland (40 degrees 34'42.5" S, 73 degrees 34'31.4" W), Chile. *Boletin de la Sociedad de Biologia de Concepcion*, *76-77*, 39-55.

Zielke, M., Solheim, B., Spjelkavik, S. & Olsen, R. A. (2005). Nitrogen fixation in the high arctic: Role of vegetation and environmental conditions. *Arctic Antarctic & Alpine Research*, *37*, 372-378.

In: Moss
Editor: Jorma Mohamed
ISBN: 978-1-63117-396-7

Chapter 2

TERRESTRIAL MOSSES AS TRACE ELEMENT BIOINDICATORS: A REVIEW

Sabina Dołęgowska*[*] *and Zdzisław M. Migaszewski
Geochemistry and the Environment Division, Institute of Chemistry,
Jan Kochanowski University, Kielce, Poland

ABSTRACT

Terrestrial mosses have been used as bioindicators since the 1960s (Rühling, Tyler, 1973). The first, large-scale studies which reported the concentrations of selected trace elements (As, Cd, Cr, Cu, Fe, Hg, Ni, Pb, V, Zn) were conducted in the Scandinavia in 1985 (Poikolainen et al., 2004). Since then many European scientists have employed the moss technique to solve various issues: (i) to document atmospheric chemistry, (ii) to map ecogeochemical changes in the environment, (iii) to pinpoint the major sources of atmospheric pollution, and (iv) to find contrast between background and "hot spot" element concentrations in polluted areas (Caritat et al., 2001).

Terrestrial mosses have found practical applications in biomonitoring studies because of their morphological and physiological properties. The lack of cuticle and root system causes that their cell walls are easily penetrated by metal ions. This enables absorption of elements and particulates from rainfall, snowmelt and dry deposition. They also have a high cation exchange capacity to retain and transport many elements.

[*] Corresponding author: e. mail: Sabina.Dolegowska@ujk.edu.pl.

Mosses are widespread, they grow in various habitats, and their annual growth increments are easily identified.

The concentrations of trace elements in moss tissues are dependent on individual moss species and their morphology, as well as bedrock lithology and many other environmental factors, i.e., topography, climate, hydrology, and edaphic (soil-related) conditions. Local topographic, climatic and edaphic variables are generally correlated with the geologic setting of study area. Different moss species show distinct variations in the concentrations of elements and some organics. This is induced primarily by diverse assimilation abilities. These facts put some constraints on the selection of a specific moss species for biomonitoring purposes. Several species (*Hylocomium splendens*, *Pleurozium schreberi*, *Scleropodium purum*, *Hypnum cupressiforme*) have successfully been used in biomonitoring programs (Berg, Steinnes, 1997), but all of these species must have acceptable sampling density, which is crucial to identify the species-dependent variability.

It should be stressed that the moss technique has many advantages. It is easier and cheaper than the conventional analysis of precipitation, which is based on a large number of sample collectors. This technique also allows us to monitor several sites at a low cost. Moreover, the element concentrations in moss tissues are high enough to make the analysis much simpler (Harmens et al., 2004).

INTRODUCTION

The control of air pollution is a very complex problem. To obtain reliable information about atmospheric conditions, sources and emissions must be identified. Moreover, the analytical methods must be adequate to pollution levels, and economical aspects must be taken under consideration. There is a lack of sensitive and inexpensive techniques to provide simultaneous monitoring of air quality. To define the spatial and temporal distribution patterns of trace elements and/or organic compounds, a long term sampling at a large number of sites is required. The use of technical equipment results from the lack of sensitive inexpensive techniques that enable the measurement of many pollutants in a large area for a long time. The EU directives on air pollution by trace elements suggest the application of biomonitors to assess a regional impact of air pollutants on the environment. The role of plant organisms in their natural habitats is significant. They are sensitive to chemical changes in the environment and they have an ability to accumulate organic and inorganic pollutants. Because the biomonitors can be used on a local and international scale the biomonitoring of air pollution has become very popular

(Prudêncio, 2006; Sabovljević et al., 2009; Marinova et al., 2010; Dołęgowska, Migaszewski, 2011; Thöni et al., 2011).

1.1. Biomonitoring and Bioindicators

The main idea of biomonitoring is to combine biological studies of living organisms with analytical techniques to assess present and future changes in the environment. This approach leads to the improvement of sensitivity and accuracy and allows us to determine elements in different biological matrices, e.g. plants or animals. Biomonitoring applies the organisms to monitor the following variables (Wolterbeek et al., 1996):

- element determinations in polluted and unpolluted areas,
- pollutant levels in the environment,
- spatial and temporal distribution of different substances.

Biomonitoring techniques are also used to predict concentrations of trace elements and/or organic compounds in the environment (Holoubek et al., 2000, 2007; Orliński, 2002; Ötvös et al., 2004; Augusto et al., 2010; Harmens et al., 2011).

Strictly speaking, biological monitoring is defined as a "response of living organisms to changes in the environment". Bioindicators indicate the bioavailability of a given pollutant over a period of time, and a qualitative determination of natural and anthropogenic factors (Burton, 1986). Biomonitoring may also be defined as a repeated measurement of organism response to chemical variability which is induced by temporal and spatial variations in the environment. This response is related to pollution levels in an examined organism and may encompass changes in species composition, and physiological and morphological structure (Osborn et al., 2000; Zhou et al., 2008; Batzias, Siontorou, 2006; Siontorou, Batzias, 2011).

Biological monitoring can be divided into two methods: active and passive (Figure 1). The passive biomonitoring includes the measurement of integrated organism exposure over a period of time, whereas the active biomonitoring encompasses the exposure of well-defined species under controlled conditions. It is interesting to note that in passive biomonitoring all processes and all sources occur at the same time and therefore there is no possibility of separating them. However, this allows us to monitor organism responses in their natural habitats (Čeburnis, Valiulis, 1999; Szczepaniak, Biziuk, 2003; Mariet et al., 2011; Rivera et. al., 2011; Ares et al., 2012; Saitanis et al., 2013).

The active biomonitoring includes: transplantation and moss bag technique. The transplantation procedure applies selected organisms (mainly mosses and lichens) and transplants them from unpolluted to polluted sites. This method defines accurately the exposure time, but it cannot assess changes in the uptake efficiency affected by variable climatic conditions The moss bag technique makes use of moss samples exposed at specific locations in permeable bags. This technique has become very popular because it can be applied in relatively high polluted areas. The exposure time is also well-defined, but the efficiency in retaining metals is strongly dependent on environmental variables. The passive technique is useful for studying large areas, whereas the active one is widely applied in areas that lack a usable native moss species (Makholm, Mladenoff, 2005; Castello, 2007; Calabrese et al., 2011).

Biomonitoring includes identification and characterization of: (i) morphological and histological changes, (ii) variations in biochemical processes, (iii) organism behavior, and (iv) population structure. It is obvious that organisms accumulate one or more elements from the environment, but all of them show a specific or unspecific effect in response to different substances (Markert, 2007; Saitanis et al., 2013). A bioindicator is an organism or a part of organism that allows us to obtain information about the state of the environment, whereas biomonitor is an organism that contains information on quantitative assessment of a degree of pollution. In the biological method we can distinguish: bioanalytics that uses the biosensors and/or biotests for environmental analyses, and biomonitoring that applies living organisms for quantitative and qualitative assessment of air, soil and water quality (Gadzała-Kopciuch et al., 2004). Biomonitor organisms provide continuous, real-time analytical information about environment quality and environmental conditions.

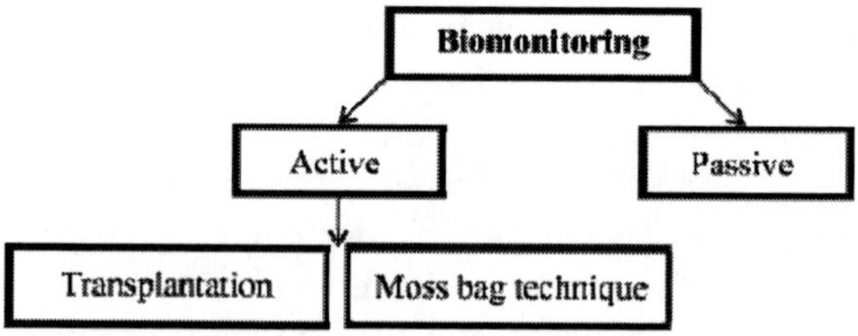

Figure 1. Biomonitoring methods.

Bioindicators may response to morphological and physiological changes in a visible way or by contrast, in an invisible way. They reflect external conditions over a certain period of time, so they are very useful in case when levels of monitoring substances change rapidly with time. In biomonitoring studies various organisms and biomaterials are used to obtain information about environment quality. Each good bioindicator must possess the following features (Conti, Cecchetti, 2001; Migaszewski et al., 2001; Chakrabortty, Paratkar, 2006):

- common occurrence – acceptable sampling density is crucial to identify the species-dependent variability,
- fast, unequivocal and distinct response to toxicants,
- tolerance of pollutants at specific levels.
- ability to assess environmental pollution over a short time (narrow tolerance to the factor examined),
- availability during all seasons,
- easiness to collect,
- high element concentrations in tissues that make the chemical analysis simpler,

In addition, the other factors should be considered:

- local conditions should not influence element uptake,
- biological changes and variations of the monitor organisms must be limited – it is necessary to predict organism responses to different conditions,
- absence of elements derived from non-atmospheric pollution sources that could make their identification difficult,
- clear and understandable mechanism of element uptake,
- low background concentrations of elements in monitor organisms,
- well-defined sampling method and sample preparation.

Living organisms have developed defense and adaptation mechanisms to survive under changing environmental conditions. They show a distinctive biological effect on environmental stress and for this reason they are so widely used in biomonitoring studies (Gadzała-Kopciuch et al., 2004; Samecka-Cymerman et al., 2006). The advantages and disadvantages of plant biomonitoring are summarized in Table 1.

Table 1. Advantages and disadvantages of plant biomonitoring

Advantages	Disadvantages
Plant analyses are cheaper than traditional techniques based on mechanical collectors.	–
Lower plants (e.g. lichens, mosses) are good bioindicators of trace elements and organic compounds.	Higher plants take up essential elements from soil and atmosphere. This provides difficulties in pollution source identification.
Plant biomonitoring allows us to provide qualitative and quantitative analysis.	Traditional, direct methods give more precise results.
Plants can be used for continuous and integrated monitoring.	It can be used mainly in retrospective studies. Short-term changes are difficult to observe.
Plants can be used for a local scale, but also for longe-range monitoring.	In comparative studies it is necessary to select the same species. Different species reveal different bioaccumulative properties.
They are widespread and occur in various habitats.	Large dispersion in a given area can reduce the selection of the species.
The results are more reliable than traditional deposition rate assessment.	In general, the lack of certified reference materials.
–	Plants show seasonal and climatic variations and are susceptible to stress.

1.2. Biomonitoring of Air Quality

Biomonitoring of air quality is an experimental method that measures the response of organisms to air pollution. Results derived from direct chemical analyses of samples taken from selected abiotic environmental compartments do not provide adequate information regarding bioavailability and potential detrimental effects of pollutants to the biosphere. Biomonitoring enables us to detect and determine trace amounts of elements and organic compounds in the environment. It is especially useful for examination of large areas. The control of air pollution includes: identification of sources and emissions as well as migration and accumulation pathways of individual chemical substances. The levels of atmospheric emissions are regulated by the law, but on the other hand many pollutants derived from unidentified sources (mostly low emissions from domestic stoves and furnaces) are released to the atmosphere (Poikolainen, 2004).

The traditional model based on collectors requires expensive equipment. Moreover, this does not provide pertinent information about temporal changes brought about by direct or indirect geochemical interactions between environmental compartments. Instrumental measurements of atmospheric deposition are not sensitive enough to pinpoint simultaneous changes in the environment. The number of observation points is limited and measurements are usually conducted in restricted areas or attributed to assess the impact of selected pollution sources. (Wolterbeek, 2002). Moreover, to obtain reliable information about air pollutant fate, transport modeling or field emission, biomonitoring should be performed for a long time.

Biomonitoring may be not only efficient, but also constitute a replacement for traditional methods using technical collectors. It allows us to provide inexpensive, large-scaled and multiple-sites monitoring. What is more, air quality bioindicators can be selected based on the ability of accumulation of airborne elements. This is the reason why widespread non-vascular lichens and mosses are commonly used to exam the atmospheric deposition. Especially noteworthy is the application of moss and lichen species which occur in nearly all climatic zones throughout the world. The results of biomonitoring studies showed that non-vascular plants may accumulate higher concentrations of trace elements compared to their vascular equivalents (Chiarenzelli et al., 2001; Uğur et al., 2004; Smagunova et al., 2004).

1.3. Biomonitoring – Summary

Biomonitoring has also found applications in epidemiological studies that deal with relationships between health effects and concentrations of trace elements and/or organic compounds in different environmental compartments. In biomonitoring studies we have to be aware of the fact that all organisms and pollutants interact with different ecosystem elements, and that organism activity is affected by various biotic and abiotic factors that occur in the environment. In addition, chemical elements may interact with one another showing synergistic or antagonistic effects. The most difficult problem regarding monitoring with living organisms is to determine the actual state of the environment and its influence on element accumulation efficiency (Markert, 2007). To obtain reliable information about organism response and potential hazard to human health, it is necessary to evaluate all interactions that occur in the environment.

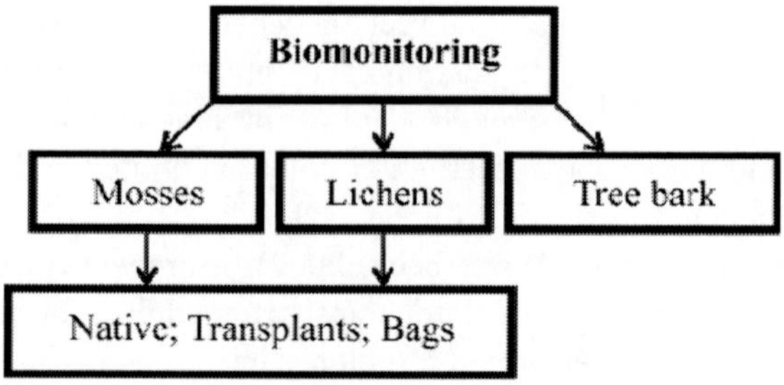

Figure 2. Scheme of biomonitoring methods using coniferous trees, lichens and mosses.

Biomonitoring studies should include: observation, sampling and measurements according to standardized procedures, as well as data handling and interpretation. At each individual step the information is obtained about the organism response, changes at the cellular level, interactions between pollution sources and health effects. Errors occurring at each step of biomonitoring studies may influence final results of measurements. Their identification requires the detailed knowledge of species, their structure, chemical composition and bioaccumulative properties, as well as characterization of environmental parameters. This approach leads to elimination of imprecise and uncertain data, and enables obtaining reliable and valuable information about environment quality. Besides, this also gives a direct insight into different substances and the way in which they are assimilated, stored and released in different conditions (Siontorou, Batzias, 2011). Currently, the use of biomonitoring techniques is especially recommended when we want to receive information on time-average concentrations of trace elements.

2. Mosses as Bioindicators of Trace Elements

In terrestrial environments many organisms or group of organisms have been adopted to monitor soil, air and water conditions and indicate the degree and extent of adverse environmental changes. Mosses, lichens and coniferous trees were found to be the most practical in biomonitoring studies. They have been successfully used to model air quality and deposition of trace elements (Onianwa, 2001; Fernández, Carballeira, 2002; Gerdol et al., 2002; Galsomiès et al., 2003; Gałuszka, 2007; Lehendorff, Schwark, 2008, Catinon et al., 2009;

Holy et al., 2009; Šakalys et al., 2009, Schröder et al., 2008). They provide information about the spatial and temporal distribution patterns of trace elements at background and "hot spot" concentrations in polluted areas. They can be used both, as passive or as active bioindicators (Figure 2) (Szczepaniak, Biziuk, 2003; Makholm, Mladenoff, 2005; Aničić et al., 2009a, 2009b; Calabrese et al., 2011; Salo et al., 2012; Saitanis et al., 2013).

2.1. Trace Element – Short Characterization

Several metals were classified by the U.S. Environmental Protection Agency as priority pollutants due to their toxicity to various organisms. Trace elements can be divided into two classes: (i) essential microelements, i.e. Cu, Fe, Mn and Zn, which are required by living organisms, and (ii) potentially toxic elements even at very low concentration, i.e. As, Cd, Cr, Hg and Pb, which are devoid of any biological function. Trace elements play an important role in physiology of living organisms, except for toxic elements, including Pb, Hg and Cd, which may lead to changes in the environment that disturb the functioning of living organisms and ultimately influence human health. They may damage or reduce mental and central nervous functions, blood composition, and lung, kidney and liver functions. Environmental exposure to toxic metals may bring about changes in DNA structure, disturbances in transcriptional regulation of oncogenes, and biocatalytic processes. Sometimes, the concentrations of essential elements may exceed critical concentrations leading to changes in catalytic and metabolic processes. Moreover, trace elements can be bioaccumulative and persistent in the environment, and this is the main reason why their concentrations must be monitored in different media (Yu, 2005; Prasad, 2008; Charlesworth et al., 2011). Based on the accumulation factor, trace elements may be divided into four groups:

1. Elements with the highest accumulation factor (10–600) – Ag, Au, Cd, Cr, Cu, Hg, Pb, Sb, Sn, Tl, Zn.
2. Elements with a high accumulation factor (1 to 10) – Ba, Bi, Ca, Fe, Mn, Mo, P, Ti, U.
3. Elements with a medium accumulation factor (0.01 to 1) – Al, As, Be, Co, Ge, K, Li, Rb, V, W.
4. Elements with a low accumulation factor (<0.01) – Ga, La, Mg, Nb, Sr, Ta, Zr.

Trace elements occur naturally in the environment and they are involved in biogeochemical cycles (Figure 3). The biogeochemical cycles of trace elements are mainly modified by human activities that induce mobilization of many elements into the biosphere and increase the amount of bioavailable chemical species in the environment. Emissions of toxic substance have a negative effect on the environment and human health. Toxicants can be transported by the air, water and living organisms and accumulate in a food chain (Bargagli et al., 2002; Gadzała-Kopciuch et al., 2004).

Trace elements are emitted into the atmosphere from natural sources such as volcanic eruptions, weathering of minerals and rocks, windblown continental dust and sea spray, as well as a result of various biological processes and anthropogenic activity. Industrial, municipal and commercial operations may release toxic and/or potentially toxic substances. A significant source of pollution is a vehicular traffic. This generates particulates, aerosols or gaseous pollutants (Carvajal et al., 2010; Rucandio et al., 2011; Ugulu et al., 2012). It should be stressed that it is relatively easy to identify air pollutants from well-defined sources compared to those released from sources devoid of specific emission outlets (low emission) (Lodenius, 2013). Trace element-rich gases and particulates released into the atmosphere undergo wet (rain, snow, fog) and dry (particles, gases) deposition (Ćujić et al., 2013). The biological monitoring gives an opportunity: (i) to study spatial distribution patterns of trace elements, (ii) to identify pollution "hot spots" and temporal trends in element concentrations, and (iii) to estimate quantitatively dry, wet and total deposition. This technique is the best method to control and predict levels of toxic elements on a local scale.

Trace elements introduced to the environment can be transported over long distances and undergo several biological and chemical processes. For example, lithophilic elements are mainly transported as wind-blown soil particles. The spatial dispersion of elements depends on their specific gravity along with the wind speed and windrose, whereas trace element exchange between plant tissues and atmosphere depends on physical, chemical, geographical and biological factors. Trace elements may be stored in biomass, leached by throughfall, transferred directly or as a litterfall to other organisms, or evaporated. Modeling of trace element dispersion and deposition is mainly based on the knowledge regarding localization of emission sources and measurements of actual atmospheric levels of trace elements. This approach is commonly used in environmental studies and is based on meteorological observations, spatial and temporal distribution and deposition of particulates, and emission records.

2.2. Mosses as Bioindicators

The idea of using mosses as bioindicators of atmospheric deposition was developed in the 1960s by Rühling and Tyler (1970, 1973). Since then, this method has further been improved and standardized, and the optimization of sampling and chemical analysis has also been investigated. The moss technique has been used for many years to assess the air quality and determine concentrations of heavy metals in atmospheric deposition (Andersen et al., 1978; Sloof, 1995; Andronikov et al., 2000; Aboal et al., 2001; Amblard-Gross et al., 2004; Harmens et al., 2004; Huttunen et al., 2005; Gałuszka, 2007; Holy et al., 2009; Migaszewski et al., 2009; Dołęgowska, Migaszewski, 2011; Rucandio et al., 2011; Thöni et al., 2011; Dołęgowska, Migaszewski, 2013).

Mosses have many properties that make them good biomarkers of air quality. Firstly, they are perennial without deciduous periods, so the biomonitoring studies can be carried out throughout the year. The lack of cuticle causes that their cell walls are easily penetrated by metal ions, and enables an easy exchange between atmosphere and cell walls. Mosses do not have roots, so they cannot take up elements from soil. This limits the absorption of elements and particulates by rainfall, snowmelt and dry deposition. This phylogenetic group is widespread, and grows in various habitats, and their annual growth increments are easily identified. They are evergreen and their tissues reach an age of at least 3-4 years. In contrast to samples collected in the traditional way, trace elements occur in the moss tissues at high concentrations. It allows us to avoid problems with inadequate limits of detection and contamination which may occur during sampling, handling and chemical analysis. The other studies have also indicated that air pollutants may be accumulated in moss tissues over a long time due to their slow growth rate and that the mosses are better adsorbents than vascular plants (Steinnes, 1995; Čerbunis, Valiulis, 1999; Fernández et al., 2000; Harmens et al., 2004; Chakrabortty, Paratkar, 2006; Salo et al., 2012; Ćujić et al., 2013; Dołęgowska et al., 2013; Lodenius, 2013). Moreover, these authors suggested that mosses showed practically neglected changes in morphology during moss lifetime, which explains their bioaccumulative properties. However, It is an arguable issue that needs further studying.

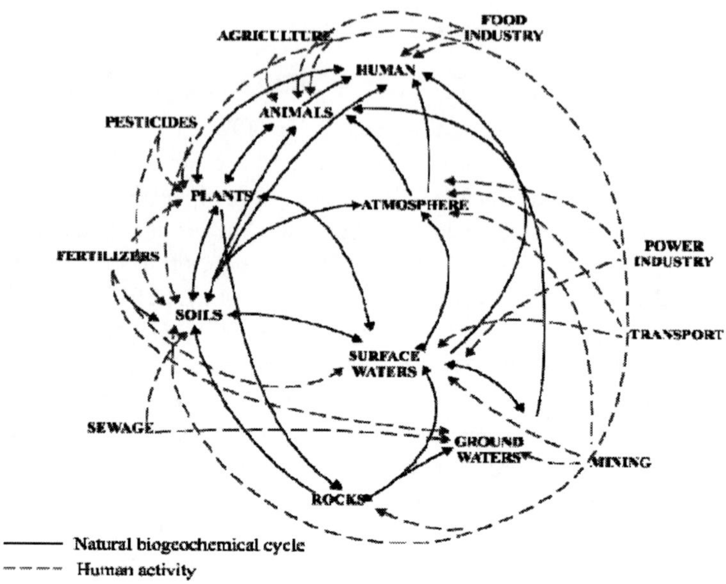

Figure 2. Biogeochemical cycle and sources of trace elements (Kabata-Pendias, Pendias, 1999; modified by the authors).

The different moss species have been used to assess the atmosphere quality and spatial and temporal distribution of trace elements. Carpet-forming species such as: *Hylocomium splendens*, *Pleurozium schreberi*, *Hypnum cupressiforme*, *Scleropodium purum* (*Pseudoscleropodium purum*) and *Abietinella abientina* have successfully been used as biomonitors of trace elements in Europe for over fifty years (Berg, Steinnes, 1997; Halleraker et al., 1998; Caritat et al., 2001; Grodzińska, Szarek-Łukaszewska, 2001; Ermakova et al., 2004; Poikolainen el at., 2004; Pesch, Schröder, 2006; Schröder et al., 2008; Barandovski et al., 2008; Migaszewski et al., 2009; Schröder, Pesch, 2010; Dołęgowska, Migaszewski, 2013; Saitanis et al., 2013). These species have many physiological and morphological properties that improves their metal uptake efficiency (Table 2). All biomonitors should be highlighted by their bioaccumulative and time-integrative abilities. The most important criteria for biomonitoring species are as follows: specificity, adsorption by dry and wet deposition, accumulation ratio, and occurrence (species richness – many authors suggest that the most widespread species be collected to maximize the sample quality and representativeness) (Fernández et al., 2000; Walterbeek, 2002). Moreover, the impact of atmospheric pollutants on biomonitor organisms must be well-known (dose-response relationship). This relationship may change over a period of time but is still very useful to characterize organism response (Garty et al., 1993). This also shows a correlation between trace element levels in organism and the distance to element sources. However, the information about changes in moss vegetation or dose-effect relationships is still limited.

From the viewpoint of biomonitoring studies, the most important is the knowledge on the processes that determine accumulation, retention and release of elements. The element concentrations in moss tissues are generally a function of element content in the air. High surface/mass ratio causes that mosses are very effective in trapping airborne soil particles. Sometimes a high metal adsorption potential may be used to remove elements from water and soil (Bargagli et al., 2002). The element accumulation process is very complex. This includes many interactions between chelating factors that occur during transport. They control the rate of trace element uptake and storage. Because all these processes affect the element efficiency during biomonitoring studies, therefore their mechanism should be better recognized.

Airborne pollutants are deposited onto moss tissues as aqueous solution (wet deposition) and gases or solid particulates (dry deposition). The metal retention is induced by various factors, i.e. layers of particulates, particulate entrapment on the surface of the cells, ion exchange, and/or passive or active cellular uptake. The entrapment of particulates is dependent on their particulate size and the tissue structure. Ion exchange is a physiological-chemical process affected by the type of free cations, cell age, desiccation process, climatic conditions (temperature, precipitation), pH and the type of pollutants. Metal cations may be bound to functional organic groups that occur in the cell walls mainly through chelatation. The elements bound to the cell walls or accumulated on the moss tissues may be leached, but the strength of this process is dependent on the binding type (Nieboer, Richardson, 1980; Brown, Brown, 1990; Zechmeister, 1995, 1998; Chakrabortty, Paratkar, 2006).

The visible changes caused by metal accumulation effect are scarce, and the most common changes occur on the cellular level. If we want to look into adsorption mechanism, we should take into account intracellular adsorption and exchange processes. Intracellular accumulation and entrapment of dust and soil particles enrich moss tissues in trace elements. For example, the trace element trend (Pb>>Zn>Cd~Cu~Cr), which is noted for *Evernia prunastri*, shows a good correlation between the metal bond and the enthalpy value in the adsorption process (Conti, Cecchetti, 2001).

Table 2. Characteristics of moss species used in biomonitoring studies (Ross et al., 1998; Berger et al., 1999)

Type of species	Morphology	Habitat
Hylocomium splendens	It is glossy, bright green, robust moss that has regularly arranged, branched branches (bipinnate). Shoots are several to 10 or even 20 cm long, which are a consequence of varying growth rates during individual seasons. It has quite large leaves about 2.5–3 mm long, and an oval, long, tapering tip. Branch leaves are smaller – about 1 mm long and have a shorter tip. Both types of leaves have a short, usually double nerve and sharply toothed margins in the upper half. The stems and branches appear furry owing to the presence of branched filaments.	*Hylocomium splendens* occurs widely in the boreal forests even in relatively harsh northern latitudes. It often occurs with moderately dense canopy featuring a forest floor of feathermosses.
Pleurozium schreberi	It has loosely pinnate branches with green leaves and bright red stem. When dry, the red stems only become visible on wetting, or after scraping leaves off the stem. Shoots are several centimeters long. The leaves are about 2–2.5 mm long; they are oval with a broad, blunt tip and envelope the stem. The stem and branch leaves are similar in shape, but the branch leaves are rather smaller. The leaves usually have a short, double nerve and thick-walled, orange-brown cells in the basal corners.	*Pleurozium schreberi* is most commonly found amongst grass and heather on heathland and in open, heathy woods. It avoids calcareous or base-rich host rocks.
Psudoscleropodium (Scleropodium) purum	It is a robust, green or yellow-green plant with more or less regularly pinnate shoots. Shoots are several to 10 cm or even more long with relatively short branches. The leaves are loosely appressed and overlapping. The leaves are about 2 mm long, broadly rounded or broadly pointed,	*Pseudoscleropodium purum* occurs in a wide range of habitats, in acidic grassland and heaths, but also in chalk and limestone grassland, on banks, among rocks and on rock

Type of species	Morphology	Habitat
	and their most distinctive feature is the presence of a small, recurved point at the tip. At the tip of the stem and new branches, the crowded points of the leaves protrude like a miniature crown. The leaves have a single nerve. Branch leaves are similar to the stem leaves, but a little smaller.	ledges. It also may occur in open woodland, but not in deep shade.
Hypnum cupressiforme	It is a widespread and abundant moss that has irregularly branched, slender to medium-sized shoots typically about 2 cm long. Leaves are strongly curved and are about 1–2 mm long. The nerve is absent, or very short and double. The shoots are green or tinged with a warm, brownish color within older parts.	*Hypnum cupressiforme* is very common on acidic to slightly base-rich bark and siliceous host rocks.
Abietinella abientina	It is green, or dark brown with a yellow or green tip shoots. Shoots are once-pinnate, with branches in 4 rows in 2 planes. Secondary stems are 5–10 cm long. Stem leaves are broadly oval, longitudinally ridged, with a broad base and tapering to a long tip. The nerve reaches about three-quarters of the way toward the tip. The branch leaves are much smaller than the stem leaves, concave, and broadly oval to spear-shaped. They have a blunt or sharp tip and the nerve extends about two-thirds of the way toward the tip.	*Abietinella abietina* prefers shallow soils in grasslands overlying chalk or limestone, calcareous sandy soils, dune holes and quarry walls. It rarely occurs on base-rich slopes and rock ledges in the mountains.

Moss accumulation tracers belong to passive processes, but the adsorbed elements may be included in plant metabolism. Many authors have assumed that the enrichment of trace elements observed in moss tissue is mainly related to the entrapment of relatively insoluble dust or soil particles. It should also be emphasized that this process may be linked to retention soluble elements in moss tissues. The study conducted by Brown and Brown (1990) showed that the accumulated metals may be transferred into the cell protoplasts and from old to young tissues, so the concentration of elements is also regulated by physiological and genetic factors. These two parameters represent a true "roulette" because chance can have a major influence.

2.3. Factors Affecting Accumulation

The trace element bioavailability and accumulation depend on many factors: moss characteristics and its morphological and physiological structure (species, age, health, type of reproduction), availability of elements (amount of nutrients), climatic (temperature, insolation, windrose, pressure systems, moisture, amount of precipitation, throughfall), topographic (elevation, aspect), geologic setting (mostly bedrock lithology), edaphic (soil moisture, texture, pH etc.) and hydrologic conditions. Local topographic, climatic and edaphic variables are generally correlated with the geologic setting of a study area (Migaszewski et al., 2001). Needless to say that sampling strategy and applied analytical methods also affect the accuracy of results obtained (Kosior et al., 2010; Bouqete et al., 2011b; Dołęgowska et al., 2013).

The climatic factors have an influence on trace element levels, but their role is still unclear. The changes in trace element concentrations induced by seasonal changes were described in detail by Markert and Weckert (1989), Ernst (1995). Wotlerbeek (2002) confirmed that natural variables in acidity, temperature, humidity, light and altitude or element bioavailibility may cause differences in bioaccumulation process. The increment of moss biomass, different growth rates and rainfall may also lead to changes in trace element uptake and accumulation in moss tissues. Moreover, the growth of moss tissues may be disturbed by metal phytotoxic effects. The accumulation process comes down to interactions between chemical species and/or between a given organism and their biotic and abiotic environment.

The concentrations of trace elements in moss samples and the efficiency uptake is dependent on chemical composition of pollutants and the type of element. Berg and Steinnes (1997) showed that the uptake efficiency differ between elements. Moreover, the sequence of trace elements depends on a moss species. The metal uptake efficiency is connected with the affinity of metals for extracellular binding sites. In *Scleropodium purum* Leblond and co-authors (2004) obtained the following sequence: Al>Cu>Zn>Pb>Hg>Fe>Ni, but Pb and Hg were the most accumulated. Zechmeister and co-authors (2003) noted in turn an enrichment in lead, but depletion in zinc and arsenic: Pb>Co, Cr>Cu, Cd, Mo, Ni, V>Zn>As. In the samples of *H. splendens* and *P. schreberi* from Poland (Dołęgowska, Migaszewski, 2013; Dołęgowska et al., 2013) the trace element concentrations decreased in the following order: Fe>Mn>Zn>Pb>Cu>Ni~Cr>Co>Hg, whereas Dragović and Mihailović (2009) found the following sequence in Serbian mosses: Mn>Zn>Ni>Pb>Cr>Cu>Cd. Ross (1990) and Chakrabortty and Paratkar (2006) indicated that higher concentrations of Pb, Cd or Co were induced by their high uptake efficiency from wet deposition. The biomonitoring studies conducted in Romania by State and co-authors (2012) revealed a statistically significant correlation between distribution of the Cr, Cu, Fe, Ni, Pb and Zn in *H. splendens*. These authors noted similar contamination factors (CF) for Cu, Fe, Ni and Pb, and the following trace element sequence: Fe>Cr>Zn, Pb>Cu>Ni. In the Salo and co-authors (2012) study the moss *S. papillosum* showed the following sequence: Fe>Al>Ti>Mn>Cr>Zn>Ba>Cu>V>Pb for samples collected in parks, and Fe>Al>Mn>Ti>V>Zn>Cu>Ba>Cr>Pb for samples collected near major roads.

The contamination factor (CF) is defined as a degree of contamination. It may be calculated as follows (1):

$$CF = \frac{c_m}{c_b} \quad (1)$$

Where the c_m is the maximum value of element concentration measured in the area and c_b is the background concentration. For $1 \leq CF < 2$ there is no contamination, for $2 \leq CF < 3.5$ there is low contamination, for $3.5 \leq CF < 8$ moderate contamination, for $8 \leq CF < 27$ severe contamination, and for $CF > 27$ extreme contamination (State et al., 2012).

The biomonitoring moss studies scarcely include determinations of major elements. Some authors (Ćujić et al., 2013) suggested that alkali and alkaline earth metals were not quantitatively retained in moss tissues. Because the uptake efficiency is strongly influenced by microenvironmental conditions and chemical properties of these metals are probably easily leached out and their concentrations in moss tissues do not reflect those in the atmosphere.

Except for the uptake efficiency for each metal by a moss species, the annual biomass production and the growth rate should be assessed. The formula for calculation of annual biomass production was suggested by Zechmeister in 1998 (2). He calculated the annual biomass production and noted that *H. splendens* exhibited the smallest annual increases per area compared to *P. schreberi*, *H. cupressiforme* and *A. abietina*, whereas *H. cupressiforme* showed the most diverse growth.

$$\boldsymbol{deposition} = \frac{concentration \cdot biomass\ production}{efficiency\ factor} \qquad (2)$$

The growth and biomass increase strongly affect the decrease of element concentrations. The moss growth is irregular and depends on availability of water, nutrients, temperature or light. The higher biomass production during the winter may lead to lower element concentrations. For example, Central European moss populations showed larger annual growth rates compared to Scandinavian mosses. It may be explained by lower mean annual temperature in Central European countries. The previous studies showed that growth rates strongly correlated with temperature and moisture but there is no significant correlation between the amount of precipitation and moss growth (Zechmeister et al., 2003).

The highest element requirements are observed during vegetation seasons, so theoretically the highest concentrations should be found in new shoots. However, Fernandez and co-authors (2000) found lower trace element concentrations in the new shoots compared to the basal parts. It may be explained by the different time of exposition. Leblond and co-authors (2004) noted statistically significant differences between apical, basal parts and a whole plant for each element, except for Na and Ca. The higher concentrations of Al, Li, Si, V, Fe, Ba, Hg and Pb were recorded in the basal parts, whereas Na, P, Ca, Mn and Zn predominated in the apical parts. The enrichment observed for the basal parts may be explained by cell degradation, leaching from soil and dust particles and the time of exposure. That is why the older moss parts show higher metal concentrations compared to the young ones For this reason, the long-term moss biomonitoring requires sampling of only one year, green, apical parts of mosses. The deposition rates are estimated as quantitative retention in apical parts. The higher concentrations of trace elements in older parts may be also elucidated by overlapping annual concentrations.

Unfortunately, currently there is no sufficient data about the growth rates and the annual production of moss species per area (Ford et al., 1995; Zechmeister, 1995; Caritat et al., 2001; Gerdol et al., 2002; Lodenius, 2013).

The moss technique is based on the assumption that trace elements come directly from the atmosphere, but in 1994 Mäkinen noted that local geology may affect moss chemistry because the elements may be adsorbed from local dust sources. Gałuszka (2007), and Dołęgowska and Migaszewski (2013) suggested that higher concentrations of Al, Sc and La in moss tissues may be linked to an enrichment from deposited soil and dust particles. The studies conducted by Zechmeister and co-authors (2003) also showed that the local site conditions and local microclimate have an important effect on moss chemistry, whereas the microclimate is dependent on the small-scale variations of the topographic pattern and the changes in deposition of heavy metals. Fernández and Carballeira (2002) noted that the moss samples collected from the areas composed of schists exhibited higher concentrations of trace elements compared to those collected from the granitic areas.

In *Scleropodium purum* the most distinct differences were recorded for Al, Co, Cr, Fe, Ni and Mn, whereas in *Hypnum cupressiforme* for As, Mn, Co, Cr and Ni. The chemistry of these moss samples was clearly dependent on different lithology. The influence of bedrock is more highlighted when comparing the element compositions of mosses from areas of different geologic setting and air pollution.

The results of comparative moss studies conducted in Wrangell–Saint Elias National Park and Preserve (Alaska) and Denali National Park and Preserve (Alaska) and the Holy Cross Mountains (Poland) exhibited that the element distribution patterns in the mosses of DENA and WSENPP are primarily affected by the geologic makeup, especially by the predominance of igneous and metamorphic rocks, as well as glacial and fluvioglacial deposits (Migaszewski et al.,2009). These deposits and soils are naturally enriched in most of the elements found in the mosses (Crock et al., 1992, 1993). In contrast, HCM mosses reflect the influence of both anthropogenic and geologic sources.

The main source of most trace elements is residential and industrial combustion of hard coal, and to a lesser extent road traffic (Migaszewski et al., 2009).

To avoid the problem arising from the geological setting, many authors suggested collecting several subsamples within a study area (Aboal et al., 2001; Aboal et al., 2006).

The correlation between lithology and moss chemistry is observed especially for REEs and other lithophilic elements (e.g. Al). This is mainly induced by deposition of windblown mineral particulates onto moss tissues. Moreover, the REEs show significant correlation with Pb, Y and Fe, which are accumulated in moss tissues as a result of particulate uptake. The study performed by Dołęgowska and co-authors (2013) showed that in *P. schreberi* and *H. splendens* REEs, i.e. La, Ce, Nd, Pr, correlated with Ba, Pb and Y. The similar behavior of lithophilic elements was noted for *P. schreberi* from southern Moravia (Czech Republic) (Sucharová, Suchara, 2004). Bargagli and co-authors (2002) pointed up that the moss samples were enriched in lithophilic elements compared to lichen samples which were abundant in atmophilic metals, i.e. Cd, Cu, Hg, Pb, V and Zn linked to long-range transport.

The similar trend in the moss *H. cupressiforme* and the lichen *Parmelia sulcata* was found by Loppi and Bonini (2000). The process of moss tissue enrichment in lithophilic elements can be natural or induced by human activity. The mentioned last increases the mobility of elements in the environment and enables taking up elements from throughfall and/or from weathering host rocks. The moss-soil/rock particle interaction explains why the ion exchange capacity increases from the moss apices to the base (Rühling, Tyler, 2004; Tyler, 2004; Schröder et al., 2008; Dołęgowska et al., 2013).

Sampling in different seasons may lead to variations in species richness and chemical composition. However, Fernández and Carballeira, (2002) noted that there is no significant seasonal differences in Cd, Hg, K, Ni and Pb concentrations in *H. cupressiforme* and *S. purum*, whereas the differences found for As, Cu and Zn were not statistically significant. Thöni and co-authors (1996) and Berg and Steinnes (1997) did not find seasonal variations in trace element concentrations in *P. schreberi* and *H. splendens*. All these differences, which were not consistent, were attributed to diverse site conditions. The significant seasonal differences in moss chemistry were in turn recorded by Markert and Weckert (1989) and Couto and co-authors (2004). Leblond and co-authors (2004) found temporal variations for all the elements examined. These authors noted the highest concentrations of Ca, Mn, Li, Al, Si, V, Fe, Ba, Hg in summer and the lowest in winter. The only exception was Na for which this relation was reversed. Fernández and co-authors (2000) found the highest Hg concentrations after the summer. In biomonitoring studies conducted by Zechmeister and co-authors (2003) the moss species *A. abietina* showed higher concentrations of elements in August and July compared to those in September.

The lowest concentrations were noted in October. The variability of Cu, Hg, Mo, Pb and Zn between sampling periods was higher than within individual sites. The whole study period was characterized by very low precipitation and poor growth of moss tissues, which may have affected the accumulation process. Boquete and co-authors (2011a) noted that the element concentrations in the moss species *Pseudoscleropodium purum* varied greatly within short periods of time, and that the degree of error associated with the temporal variability was distinctly high.

The seasonal variability of trace element concentrations in mosses is still a controversial issue because there is no explicit information when the moss samples should be collected (Table 3) (Acar, 2006). For example, Berg and Steinnes (1997) suggested that the moss sampling campaign should be carried out during the summer season. Changing weather conditions explain significant differences of element concentrations in moss transplants. Environmental conditions, including windrose or precipitation, strongly affect the element concentrations in moss tissues used for active monitoring (Zechmeister et al., 2003).

Throughfall and stem flows are the next factors that affect trace element concentrations. The effect of throughfall in moss varies and depends on the weather conditions and position of sampling sites (between the crowns), whereas the amount of precipitation depends on the type of tree crown. The differences in deposition inside and outside the forest are determined by the behavior of the canopy. The study conducted by Fernandez and Carballeira (2002) displayed element enrichments whereas Bargagli and co-authors (2002) noted that the atmospheric deposition under the tree canopy may lead to changes in element concentrations. They indicated that the samples of *H. cupressiforme* and *S. purum* collected from forested areas had lower concentrations of elements compared to those collected from open areas. Because there is no explicit information about the influence of throughfall on moss chemistry, the moss samples should be collected in open areas. The moss samples collected in large open areas in forests and under canopies showed differences in element contents. The concentration of such elements as: Al, As, Co, Cr, Fe, Hg, S and V decreased significantly with an increasing forest cover (Sucharová, Suchara, 2004).

The process of trace element accumulation and/or leaching is induced by the intensity and type of precipitation. This controls deposition of elements, washing out of dry deposition, leaching of elements and affecting moss production. The amount of precipitation correlates with the element bioaccumulation efficiency due to the atmospheric rain-out and wash-out effect. Accumulation through dry deposition predominates in the arid climates compared to the humid. Couto and co-authors (2004) noted that the moss samples collected after drought were enriched in lithophilic elements because during the dry seasons trace elements are translocated from one to another part of the moss. Dry conditions and droughts may lead to dusty atmosphere and increase the contents of elements derived from dry deposition. Dust and soil particles deposited onto moss tissues enrich the moss samples mainly in Fe, Cr, Al or Ti. Different climatic conditions like acid deposition or the presence of sea salts also influence accumulation processes. In dry and unpolluted areas the chemical composition of unwashed moss samples is regulated by trapped soil and rock dust, whereas in natural mineralized areas moss samples reveal the geochemical signatures of the region (Berg, Steinnes, 1997; Čerbunis, Valiulis, 1999; Bargagli et al., 2002; Couto et al., 2004).

Soil and dust particles deposited onto tissues may be also washed out by intensive rain. This brings about decreasing in element concentrations (Schröder et al., 2008). The results obtained for moss samples collected after a dry period or after rainfall should be discussed critically (Leblond et al, 2004).

2.4. Interspecies Differences

All biological organisms show differences between species. Many different moss species have been used in biomonitoring studies (see section 2.2), but no explicit information is available on which species is the best trace element bioindicator. Various moss species have different abilities to take up and retain trace elements. This is brought about by diverse ecophysiology and mechanisms of element bioaccumulation (Bargagli et al., 2002). The authors suggested that interspecies calibrations be performed due to different bioaccumulative properties of individual moss species. The chemical differences may occur between species, between individual organisms of the same species, and/or between separate parts of single specimen (see section 2.1).

The correlation between the uptake efficiency and trace element concentrations in *P. schreberi* and *S. purum* and between *P. schreberi* and *H. cupressiforme* was examined by Siewers and Herpin (1998), but the authors found no relationships between the moss species and the element contents. In the early 90s Köhler and Peichl (1993) indicated that *H. cupressiforme* showed higher bioaccumulative properties than *P. schreberi*. Higher concentrations of Ni, Pb and Zn in *H. cupressiforme* versus *P. schreberi* were also measured by

Thöni and co-authors (1996). Schröder and co-authors (2008) noted distinctly higher concentrations of As, Cd, Hg, Ni and Pb in *H. cupressiforme* than in the other moss species.

Interspecies comparison of element concentrations in *P. schreberi*, *S. purum* and *H. cupressiforme* in the Czech Republic did not reveal significant differences in Al, As, Cd, Co, Cr, Cu, Fe, Mo, Ni, Pb, S, V and Zn levels in *P. schreberi* and *S. purum*. In contrast, *H. cupressiforme* showed enrichments in all the elements examined (Suchara, Sucharová, 1998). Bargagli and co-authors (2002) also found higher concentration of Al, Fe and Ti in *H. cupressiforme*.

Normalization of trace elements to Al and Ti contents in moss samples exhibited no correlation between As, Ba, Cr, Cu, Fe, Hg, Ni, Pb, V or lithophilic elements and their distribution in the study area. Moreover, the authors suggested that *H. cupressiforme* was very tolerant for atmospheric pollutants and arid climatic conditions, and it may be used to obtain maps of trace element deposition in different areas. Fernandez and Carballeira (2002) concluded that differences in trace element concentrations between *S. purum* and *H. cupressiforme* arose from interspecific differences in bioaccumulation of selected elements such as: Al, Fe and Mn. The comparison between *P. schreberi, S. purum* and *H. cupressiforme* showed that *P. schreberi* revealed markedly higher concentrations of Al, As, Au, Ce, Cl, Cs, La, Mo, Na and Zr than *H. cupressiforme*, whereas *S. purum* higher levels of Al, Ce, Cr, Eu, Fe and Ti. Compared to those two species, *H. cupressiforme* was enriched in Ca, Cl, Ni and Ti (Galsomiès et al., 2003). The higher bioaccumulation capacity of *H. cupressiforme* was also noted by Carballeira and co-authors (2008). These authors found higher concentrations of all the elements examined (except for Al and As) in *H. cupressiforme* compared to *S. purum*. Zechmeister (1998) found that apart from Cu, *H. cupressiforme* has higher retention capacities compared to *A. abietina*, *P. scheberi* and *H. splendens*, but the uptake efficiency is similar.

Statistical comparison of bioaccumulative properties of two moss species *H. splendens* and *P. schreberi* reported by Dołęgowska and co-authors (2013) showed that *H. splendens* had higher bioaccumulative properties than *P. schreberi*. The authors found higher concentrations of Ba, Ce, Co, Cu, Fe, Hg, La, Mo, Ni, Pb and V in *H. splendens*, but similar levels of Cr, Dy, Er, Ho, Sn, Tb, Th and Yb in these two moss species.

The same trend was observed by Ross (1990), Berg and Steinnes (1997), and Halleraker and co-authors (1998). Harmens and co-authors (2004) noted similar concentrations of Zn and As in *H. splendens* and *P. schreberi*. In addition, the higher concentrations of Cr, Fe, Ni, Cu and Pb were recorded in *H. splendens*, and V and Cd in *P. schreberi*. Ćujić and co-authors (2013) assumed that diverse element concentrations in mosses from Serbia resulted from different sampling seasons and diverse moss species used in biomonitoring studies. These authors did not find significant differences in concentrations of Al, Ca, Fe, K, Mg, Mn and Si in *Brachythecium sp.* and *Kindbergia praelonga*. The highest Al levels in these moss species corresponded to a high content of this element in earth's crust, and were induced by deposition of mineral and soil particles onto moss tissues.

Castello (2007) showed that transplants of *Pseudoscleropodium purum* and *Hypnum cupressiforme* varied in the uptake of different elements. In contrast to Carballeira and co-authors (2008), Castello found higher concentrations of Al, Fe, Ti, and Pb in *P. purum*, whereas Cu and V predominated in *H. cupressiforme*. In active biomonitoring, the uptake and retention of elements is also dependent on morphological features of transplanted mosses.

The higher accumulation capacity of *P. purum* transplants suggests that in specific conditions less crowded leaves are more efficient in element retention. Tabors and co-authors (2004) also noted differences in chemical composition of *H. splendens* transplant in unpolluted and polluted areas.

A suitable indicator for assessing accumulative properties of different moss species is the element concentration ratio. The *P. schreberi/H. splendens* (*Pl/Hy*) median element concentration ratios calculated by Dołęgowska and co-authors (2013) (Table 4) are similar to those presented by Ross (1990) and Wolterbeek and co-authors (1995). Reimann and co-authors (2001) and Gałuszka (2007) received lower values for Ba, Cd, Mn and Sr, while Berg and Steinnes (1997) and Halleraker and co-authors (1998) obtained higher values for Pb and V. It is interesting to note that the same interspecies concentration distribution patterns were found for Co, Cu, Fe, Hg, Mo, Ni and Zn in the moss species from Poland.

Differences in element concentrations are the most probably dependent on the moss species morphology, leave structure, physiology, binding elements by organic compounds (chelators), environmental conditions (including pH and moisture) and the presence of other elements. Identification of interspecies differences is important when we want to compare results derived from various biomonitoring studies.

Table 3. Seasonal variations in trace element concentrations in different moss species

Author	Moss species	Sampling time	Found variations
Markert and Weckert, 1989	*Polytrichum formosum*	Bimonthly over 2 years	Seasonal variations for Al, Ba, Ca, Fe, Pb, Zn
Thöni et al., 1996	*Hylocomium splendens, Pleurozium schreberi, Hypnum cupressiforme*	Monthly – from March to August	No statistically significant variations for Al, Fe, Pb, V, Zn.
Berg and Steinnes, 1997	*Hylocomium splendens, Pleurozium schreberi*	June, July, September	No statistically significant variations for Ba, Ca, Fe, Li, Mn, Pb, V, Zn.
Fernández, Carballeira, 2002	*Scleropodium purum*	Monthly over 1 year	No statistically significant variations between four seasons.
Couto et al., 2004	*Scleropodium purum*	Monthly over 3.5 years	Seasonal variations for Na, no seasonal variations for Al, Ca, Fe, Hg, Mn and Zn.
Zechmeister et al., 2003	*Abietinella abietina*	Monthly – from July to October	No variations for a single elements, seasonal variations for all elements together.
Leblond et al., 2004	*Scleropodium purum*	Every other week over 1 year	Seasonal variations for Al, Ba, Ca, Fe, Hg, Li, Mn, Na, Si, V.

Table 4. ***Pl/Hy*** **concentration ratios of 13 elements in** ***H. splendens*** **and** ***P. schreberi*** **derived from selected moss studies**

Moss study	Median values of *Pl/Hy* element concentration ratios												
	Ba	Cd	Co	Cu	Fe	Hg	Mn	Mo	Ni	Pb	Sr	V	Zn
Dołęgowska et al., 2013	1.00	1.14	0.89	0.90	0.91	0.86	1.05	0.79	0.85	0.76	1.05	0.89	1.03
Gałuszka, 2007	0.94	1.09	0.93	0.89	0.95	0.96	0.81	0.75	0.88	0.88	1.04	0.92	0.97
Reimann et al., 2001	0.95	0.96	0.97	0.80	0.89	0.98	0.95	0.52	0.90	0.99	0.92	0.97	1.03
Halleraker et al. ,1998	1.30	1.10	0.91	0.80	0.89	0.76	1.19	0.73	0.93	1.14	1.07	0.93	1.09
Berg, Steinnes, 1997	–	1.18	0.89	0.76	0.63	–	1.30	0.68	0.53	0.90	–	1.14	–
Wolterbeek et al., 1995	–	1.14	–	0.93	0.86	–	–	–	0.78	0.86	–	0.86	1.00
Ross, 1990	–	–	–	0.84	0.85	–	–	–	0.83	0.83	–	0.79	–

Table 5. Selected data on major moss projects carried out in European countries since 1975

Authors	Country	Moss species	Year	Analytical method	Reference material	Analyzed elements
Grodzińska et al., 1999	Poland	*Pleurozium schreberi*	1975-1995	AAS	Pine Needles SRM 1575, Tomato Leaves SRM 1573a	Cd, Cr, Cu, Fe, Ni, Pb, Zn
Caritat et al., 2001	Barents Region (NE Norway, N Finland, NW Russia)	*Pleurozium schreberi, Hylocomium splendens*	-	ICP-MS, ICP-AES	Peach Leaves SRM 1547, Moss 183	Al, B, Ba, Ca, K, La, Mg, Mn, Na, P, Rb, Si, Sr, Th, U, Y
Galsomiès et al., 2003	France	*Pleurozium schreberi, Scleropodium purum, Hypnum cupressiforme*	1995-1996	INAA, ICP-MS	Lichen 336 IAEA, Standard Moss Reference Materials M1, M2, M3	Al, As, Cl, Co, Cs, Fe, La, Mn, Pb, Rb, Sb, Sc, Sm, V, Zn

Authors	Country	Moss species	Year	Analytical method	Reference material	Analyzed elements
Ermakova et al., 2004	Central Russia	*Pleurozium schreberi, Hylocomium splendens,*	2000-2002	INAA, AAS	Lichen 336 IAEA, Pine Needles SRM 1575, Moss Reference Sample NORD DK-1	Al, As, Ba, Br, Ca, Cd, Ce, Cl, Co, Cr, Cs, Cu, Fe, Hf, K, La, Mg, Mn, Mo, Na, Ni, Pb, Rb, Sb, Sc, Sm, Sr, Ta, Tb, Th, U, V, W, Zn
Harmens et al., 2004, 2008	European countries	*Pleurozium schreberi, Hylocomium splendens, Hypnum cupressiforme*	1990-2000 2000-2001	ICP-AES; ICP-MS, INAA, AFS	Standard Moss Reference Materials M3	As, Cd, Cr, Cu, Fe, Hg, Ni, Pb, V, Zn
Rühling, Tyler, 2004	Sweden	*Pleurozium schreberi*	1975-2000	ICP-AES, ICP-MS	–	Ag, Al, As, B, Ba, Be, Bi, Ca, Fe, K, Mg, Mn, Na, S, Si, Sr
Suchara, Sucharová, 1998; Sucharová, Suchara, 2004	Czech Republic	*Pleurozium schreberi, Scleropodium purum (Eurhynchium angustirete, Brachythecium rutabulum, Brachythecium salebrosum)*	1995, 2000	ICP-MS, AMA 254	Hay V 10 IAEA, Lichen 336 IAEA, Apple Leaves SRM 1515, Pine Needles SRM 1575	Ag, Al, As, Ba, Be, Bi, Cd, Ce, Co, Cr, Cs, Cu, Fe, Ga, Hg, In, La, Li, Mn, Mo, Ni, Pb, Pr, Rb, S, Sb, Sc, Se, Sr, Th, Tl, U, V, Y, Zn
Poikolainen et al., 2004, 2009	Finland	*Pleurozium schreberi, Hylocomium splendens,*	1990-2006; 1985-2000	CV-AFS, GF-AAS, Kjeldahl method	Citrus leaves SRM 1572, Tomato Leaves SRM 1573a	As, Cd, Cr, Cu, Fe, Hg, N, Ni, Pb, V, Zn
Reimann C. et al., 2006	Norway (120 km long south-north transect)	*Hylocomium splendens,*	2005	ICP-MS	Pine Needles SRM 1575, Standars V13	Ag, Al, As, Au, B, Ba, Bi, Ca, Cd, Co,

Table 5. (Continued)

Authors	Country	Moss species	Year	Analytical method	Reference material	Analyzed elements
					Mountain Hemlock Needles	Cr, Cu, Fe, Ga, Hg, K, La, Mg, Mn, Mo, Na, Ni, P, Pb, Pt, S, Sb, Sr, Te, Th, Ti, Tl, U, V, W, Zn
Schröder et al., 2008 Schröder, Pesch, 2010	European Countries (Austria, Czech Republic, Germany, Italy, Slovakia, Slovenia, Poland, Switzerland	*Pleurozium schreberi, Scleropodium purum, Hypnum cupressiforme*	1990, 1995, 2000, 2005	ICP-MS, ICP-AES, AMA 256, CV-AAS, GF-AAS, HG-AAS, INAA	Standard Moss Reference Materials M2, M3	Al, As, Cd, Cr, Cu, Fe, Hg, Ni, Pb, Sr, Ti, V, Zn
Barandovski et al., 2008	Macedonia	*Hypnum cupressiforme, Camptothecium lutescens, Homalothecium sericeum*	2002	INAA, GF-AAS, CV-AAS	Lichen 336 IAEA, Moss Reference Sample NORD DK-1	Ag, Al, As, Au, Ba, Br, Ca, Cd, Ce, Cl, Co, Cr, Cs, Cu, Eu, Fe, Hf, I, In, K, La, Mg, Mn, Mo, Na, Ni, Rb, Sb, Sc, Se, Sm, Sr, Ta, Tb, Th, Ti, U, V, W, Zn, Zr
Coşkun et al., 2009	Bulgaria, Turkey	*Hypnum cupressiforme*	2003-2005	ICP-AES	Standard moss reference materials M2, M3	Al, As, Cd, Cr, Cu, Fe, Ni, Pb, Sb, V, Zn
Holy et al., 2009	France	*Scleropodium purum, Hypnum cupressiforme, Thuidium tamariscinum, Pleurozium schreberi, Hylocomium splendens*	2006	ICP-AES, CV-AAS,	Standard moss reference materials M2, M3	Al, As, Ca, Cd, Cr, Cu, Fe, Hg, N, Na, Ni, Pb, Sb, V, Zn

Authors	**Country**	**Moss species**	**Year**	**Analytical method**	**Reference material**	**Analyzed elements**
Holy et al., 2009	France	*Scleropodium purum, Hypnum cupressiforme, Thuidium tamariscinum, Pleurozium schreberi, Hylocomium splendens*	2006	ICP-AES, CV-AAS,	Standard moss reference materials M2, M3	Al, As, Ca, Cd, Cr, Cu, Fe, Hg, N, Na, Ni, Pb, Sb, V, Zn
Šakalys et al., 2009	Lithuania, Czech Republic		1995-2005	AAS, ICP-MS, AMA 254	Pine Needles SRM 1575, Hay V-10, Sea Lettuce SRM 279, Plankton SRM 414, Standard Moss Reference Materials S3, DK2, DK3, M2, M3	As, Cd, Cr, Cu, Fe, Hg, Mn, Ni, Pb, V, Zn
Thöni et al., 2011	Bulgaria; Switzerland	*Pleurozium schreberi, Hypnum cupressiforme*	1990-2005	ICP-AES, ICP-MS	Standard moss reference materials M2, M3	Cd, Cr, Cu, Fe, Ni, Pb, V, Zn

2.5. Long-Term and Long-Rage Biomonitoring Studies

The moss technique is a part of monitoring programs for air quality in European countries that have been conducted since 1975 (Table 5). The Heavy Metals in Mosses Survey was established in 1980. The first biomonitoring program started in the Scandinavian countries in 1985 (Rühling, 1989). Since then, this program has been developed and included large parts of Europe (Caritat et al., 2001). In 1995 the moss biomonitoring programs were conducted by most north European countries (Table 3). At the beginning, these studies encompassed determinations only the most toxic heavy metals, i.e. As, Cd, Hg or Pb, but after some time the scope of research has been extended to include other elements (Migaszewski et al., 2010, Salo et al., 2012)

In Poland the first long-range study was carried out by Grodzińska and co-authors (1999). The principal objective of this study was to present the regional atmospheric deposition of trace (heavy) metals in Poland in 1995 and to compare these data with the results obtained 20 years earlier in Polish National Parks. The study was conducted as a part of Harmens's project (2004), which included determinations of ten trace elements in naturally growing mosses from twenty-eight European countries. This project encompassed almost 7000 sampling sites and its main purpose was to obtain information about the distribution patterns of trace metals in mosses across the continent, and to identify the most polluted areas and to demarcate transboundary pollution pathways.

The next major biomonitoring program was performed in the Barents Region (northeastern Norway, northern Finland and northwestern Russia) (Caritat et al., 2001; Reimann et al., 2001). This geochemical mapping project covered an area of about 188,000 km^2. The investigation began in 1992 with mapping of a subarea of about 12,000 km^2 around the Ni smelter at Nikel (Kola Peninsula, Russia). This continued during 1994 and 1995 and included regional geochemical mapping of the study area. The project also encompassed a large-scale biomonitoring programs in Sweden conducted by Rühling and Tyler (2004) and in the Czech Republic in 2000 by Sucharová and Suchara (2004). In 2002 the first biomonitoring study of air quality was carried out in the Republic of Macedonia (Barandovski et al., 2008), This allowed Macedonia to be added to the European network of atmospheric element deposition based on moss biomomitoring. During the Macedonian study, the special attention was drawn to uranium mineral deposits and the content of this element in moss tissues. The authors pinpointed uranium geochemical anomalies in some regions of Macedonia. The European moss survey showed that the moss technique is a cheap, effective and accurate to monitor deposition of trace elements and temporal trends. This also provided information about the deposition patterns of different elements. Moreover, the moss investigations allowed evaluation of air quality and determination of different pollutants throughout the continent.

For several decades the biomonitoring studies have also been conducted on a small scale to identify and assess the influence of trace element pollution sources on the local environment. For example, Fernández and Carballeira (2002) used the moss species *S. purum* to exam the area surrounding a coal-fired power plant and evaluate its impact on the environment during 1995 through 1997. Dragović and Mihailović (2009) used the moss technique to assess the contribution of emission sources to pollution of areas remote from industrial facilities at Zlatibor in Serbia.

2.6. Advantages and Disadvantages of Moss Technique

The moss biomonitoring uses statistical techniques to group and interpret the results obtained. This allows us to identify and characterize pollution sources and to indicate the "hot spots" (Marinova et al., 2010). This technique has several advantages and disadvantages compared to the traditional one (Sloof, 1995; Conti, Cecchetti, 2001; Bargagli et al., 2002; Fernández, Carballeira, 2002; Wolterbeek, 2002; Gadzała-Kopciuch et al., 2004; Harmens et al., 2004; Castello, 2007; Markert, 2007; State et al., 2012; Ugulu et al., 2012).

2.6.1. Advantages

- Monitor organisms are permanent and common in the environment. This characteristic enables us to provide a time-integrated measure of spatial trace element distribution over a period of time.
- Mosses show stability of accumulation with time, and concentrations of trace elements in moss tissues are linearly related to those in bulk deposition. Moreover, their response is usually repeatable in various conditions.
- Some species may be suitable for monitoring one element and the others for a group of elements.
- The moss samples can be collected over the whole year, thus enabling interpretation of seasonal variations.

- The higher concentrations of trace elements in moss tissues make a chemical analysis more reliable and reduce the possibility of contamination.
- The moss technique gives a feasible assessment of trace element concentrations in airborne particulates and deposition on a large scale.
- The moss technique allows us to prepare air quality maps in order to obtain information about pollution of the study area in a short time.
- The easy and fast moss technique enables selection of a greater number of sampling sites and samples providing more information on a degree of pollution. This technique is also cheaper than a conventional approach, because it does not need a large number of collectors for routine sampling and technical equipment.
- Biological responses of organisms to pollution levels in the environment are more reliable than the information obtained from chemical and physical detectors.

2.6.2. Disadvantages

- In biomonitoring studies a much more attention should be paid to climatic and environmental factors. Spatial-temporal variations may lead to sampling errors and changes in trace element bioavaibility.
- Low detection limits of trace elements bring about analytical problems. Bioindicator samples displaying very low concentrations of these elements must be treated with caution.
- Lack of methods to assess the levels of elements contained in low emissions.
- Toxic effects of trace elements on biomonitor organisms may lead to unintended bioconcentrations and/or biomagnifications.
- Diverse element tolerance limits of moss species also lead to erroneous interpretation of different organism response.
- The interspecies differences between moss species affect trace element concentrations and accumulation processes. The presence of divergence between input and output of elements is connected with many environmental factors.
- The processes affecting uptake and binding mechanisms are poorly known. Moreover, the quantitative aspect is still not well understood and the biological effects of atmospheric deposition are also unrecognizable.
- The strong influence of soil composition on moss chemistry showed that the element concentrations must be normalized to the Al and Ti to

eliminate inputs of elements from soil and rock particles deposited onto moss tissues.
- To provide temporally representative data throughout the year, a large number of samples should be collected.
- Mosses may be characterized by an irregular and patchy distribution. This is the reason why sampling should be conducted by a specialist who is capable of identifying individual species. Anthropogenic pressure may also cause a lack of monitor species in a given area.
- The moss technique is applied mainly for unpolluted areas because bioindicators may be absent in polluted areas. However, natural ecosystems are characterized by complex processes, and therefore environmental studies should always include interactions between moss species and chemical elements.
- Biomonitoring studies must be completed with information about emissions which may give essential data about chemical element forms.

CONCLUSION

Some terrestrial moss species are widespread in many climatic zones and specific environmental compartments. Moreover, they belong to the perennial phylogenetic group and therefore can monitor chemical variations in the environment throughout the year. The study conducted by Gough and Crock (1997) in Denali National Park and Preserve (Alaska) indicated that mosses are better biomonitors of airborne metals than are lichens (and both plants are considerably better than is soil). Another advantage is that mosses are easy to collect without using expensive technical equipment. The active and passive biomonitoring techniques have successfully been applied for determination of air quality. Both provide pertinent information about spatial distribution patterns and critical levels of trace elements. The terrestrial mosses have also found other applications: (i) to monitor temporal and spatial changes in the environment, (ii) to identify emission sources, (iii) to track the pathways of pollutants, and (iv) to pinpoint the "hot spots" of toxicants that may be linked to geochemical anomalies generated by contemporary and historical mining and industrial activities, urbanization, traffic emissions, long-range transport etc. Moss bioindicators give information about ecological and health hazards. However, more data are needed about physiological and biochemical processes in mosses that may influence trace element concentrations.

The lack of uniform and consistent methodology usually prevents from comparison of data derived from different studies. This is the reason why the most crucial issue in biomonitoring moss studies is to work out standardized protocols for survey planning, sampling strategy, sample preparation, chemical analysis and data handling (Steinnes, 1995). The biomonitoring procedures must consider the following major issues: (i) to get to use the advantages and possibilities of biomonitoring, (ii) to reduce uncertainty of biological monitoring, and (iii) to define biotic and abiotic parameters needed for proper data interpretation. The universal scheme worked out for the moss technique enables the comparison of results from different regions. Moreover, this can be modified for a given area by taking into account its specific climatic and environmental characteristics. The state of knowledge about all factors that influence accumulation of trace elements allows us to obtain reliable results and find correlations between trace element concentrations in moss tissues and atmospheric deposition.

REFERENCES

Aboal, J. R., Couto, J. A., Fernandez, J. A. & Carballeira, A. (2006). Definition of number of subsamples for using mosses as biomonitors of airborne trace elements. *Arch. Environ. Contam. Toxicol.*, *50*, 88-96.

Aboal, J. R., Fernandez, J. A. & Carballeira, A. (2001). Sampling optimization, at site scale, in contamination monitoring with moss, pine and oak. *Environ. Pollut.*, *115*, 313-316.

Acar, O. Biomonitoring and annual variability of heavy metal concentration changes using moss. (2006). (*Hypnum cupressiforme* L. ex. Hedw.) in Canakkale Province. *J. Biol. Sci.*, *6*, 38-44.

Amblard-Gross, G., Maul, A., Férard, J. -F., Carrto, F. & Ayrault, S. (2004). Spatial variability of sampling: grid size impact on atmospheric metals and trace elements deposition mapping mosses. *J. Atmos. Chem.*, *49*, 39-52.

Andersen, A., Hovmand, M. & Johnsen, I. B. (1978). Atmospheric heavy metal deposition in the Copenhagen area. *Environ. Pollut.*, *17*, 133-151.

Andronikov, S. V., Davidson, D. A. & Spiers, R. B. (2000). Variability in contamination by heavy metals: sampling implications. *Water, Air, Soil Pollut.*, *120*, 29-45.

Aničić, M., Tasić, M., Frontasyeva, M. V., Tomašević, M., Rajšić, S., Strelkova, L. P., Popović, A. & Steinnes, E. (2009a). Active biomonitoring with wet and dry moss: a case study in an urban area. *Environ. Chem. Lett.*, *7*, 55-60.

Aničić, M., Tomašević, M., Tasić, M., Rajšić, S., Popović, A., Frontasyeva, M. V., Lierhagen, S. & Steinnes, E. (2009b). Monitoring of trace element atmospheric deposition using dry and wet moss bags: accumulation capacity versus exposure time. *J. Hazard. Mater.*, *171*, 182-8.

Ares, A., Aboal, J. R., Carballeira, A., Giordano, S., Adamo, P. & Fernández, J. A. (2012). Moss bag biomonitoring: A methodological review. *Sci. Total Environ.*, *432*, 143-158.

Augusto, S., Máguas, C., Matos, J., Pereira, M. & Branquinho, C. (2010). Lichens as an integrating tool for monitoring PAH atmospheric deposition: A comparison with soil, air and pine needles. *Environ. Pollut.*, *158*, 483-489.

Barandovski, L., Cekova, M., Frontsayeva, M. V., Pavlov, S. S., Stafilov, T., Steinnes, E. & Urumov, V. (2008). Atmospheric deposition of trace element pollutants in Macedonia studied by the moss biomonitoring technique. *Environ. Monit. Assess.*, *138*, 107-118.

Bargagli, R., Monaci, F., Borghini, F., Bravi, F. & Agnorelli, C. (2002). Mosses and lichens as biomonitors of trace metals. A comparison study on *Hypnum cupressiforme* and *Parmelia caperata* in a former mining district in Italy. *Environ. Pollut.*, *116*, 279-287.

Batzias, F. A. & Siontorou, C. G. (2006). A knowledge-based approach to environmental biomonitoring. *Environ. Monit. Assess.*, *123*, 167–197.

Berg, T. & Steinnes, E. (1997). Use of mosses (*Hylocomium splendens* and *Pleurozium schreberi*) as biomonitors of heavy metal deposition: from relative to absolute deposition values. *Environ. Pollut.*, *98*, 61-71.

Berger, T. E., De Pontieu, B., Fletcher, L., Schrijver, C. J., Tarbell, T. D. & Title, A. M. (1999). What is moss? *Sol. Phys.*, *190*, 409-418.

Boquete, M. T., Fernández, J. A., Aboal, J. R. & Carballeira, A. (2011a). Analysis of temporal variability in the concentrations of some elements in the terrestrial moss *Pseudoscleropodium purum*. *Environ. Exp. Bot.*, *72*, 210-216.

Boquete, M. T., Fernández, J. A., Aboal, J. R. & Carballeira, A. (2011b). Are terrestrial mosses good biomonitors of atmospheric deposition of Mn? *Atmos. Environ.*, *45*, 2704-2710.

Brown, D. H. & Brown, R. M. (1990). Reproducibility of sampling for element analysis using bryophytes. In *Element concentration cadasters in ecosystems*, Lieth, H.; Markert B. Eds.; VCH Publishers: Weinheim, 55-62.

Burton, M. A. S. (1986). *Biological monitoring of environmental contamination (plants)*; MARC Technical Report: King's College London.

Calabrese, S., D'Alessandro, W., Parello, F., Bellomo, S. & Brisca, L. *Books of Abstracts*, Application of the moss bag biomonitoring technique in an active volcanic environment (Mt. Etna, Italy). 11th Gas Workshop, Kamchatka, Russia, September 1-10, 2011; Commission on the Chemistry of Volcanic Gases (CCVG) – IAVCEI.

Carballeira, C. B., Aboal, J. R., Fernández, J. A. & Carballeira, A. (2008). Comparison of the accumulation of elements in two terrestrial moss species. *Atmos. Environ.*, *42*, 4904-4917.

Caritat, P., Reimann, C., Bogatyrev, I., Chekushin, V., Finne, T. E., Halleraker, J. H., Kashulina, G., Niskavaara, H., Pavlov, V. & Äyräs, M. (2001). Regional distribution of Al, B, Ba, Ca, K, La, Mg, Mn, Na, P, Rb, Si, Sr, Th, U and Y in terrestrial moss within a 188,000 km^2 area of the central Barents region: influence of geoecology, seaspray and human activity. *Appl. Geochem.*, *16*, 137-159.

Carvajal, B., Aboal, J. R., Fernández, J. A., Real, C. & Carballeira, A. (2010). Influence of roads and inhabited areas on metal concentrations in terrestrial mosses. *Atmos. Environ.*, *44*, 3432-3441.

Castello, M. (2007). A comparison between two moss species used as transplants for airborne trace element biomonitoring in NE Italy. *Environ. Monit. Assess.*, *133*, 267-276.

Catinon, M., Ayrault, S., Clocchiatti, P., Boudouma, O., Asta, J., Tissut, M. & Ravanel, P. (2009). The anthropogenic atmospheric elements fraction: a new interpretation of elemental deposits tree barks. *Atmos. Environ.*, *43*, 1124-1130.

Čerbunis, D. & Valiulis, D. (1999). Investigation of absolute metal uptake efficiency from precipitation in moss. *Sci. Total Environ.*, 226, 247-253.

Chakrabortty, S. & Paratkar, G. T. (2006). Biomonitoring of trace element air pollution using mosses. *Aerosol Air Qual. Res.*, *6*, 247-258.

Charlesworth, S., De Miguel, E. & Ordóñez, A. (2011). A review of distribution of particulate trace elements in urban terrestrial environments and its application to consideration of risk. *Environ. Geochem. Hlth.*, *33*, 61-71.

Chiarenzelli, J., Aspler, L., Dunn, C., Cousens, B., Ozarko, D. & Powis, K. (2001). Multi-element and rare earth element composition of lichens, mosses, and vascular plants from Central Barrenlands, Nunavut, Canada. *Appl. Geochem.*, *16*, 245-270.

Conti, M. E. & Cechcetti, G. (2001). Biological monitoring: lichens as bioindicators of air pollution assessment – a review. *Environ. Pollut.*, *114*, 471-492.

Coşkun, M., Yurukova, L., Çayir, A., Coşkun, M. & Gecheva, G. (2009). Cross-border response of mosses to heavy metal atmospheric deposition in Southeastern Bulgaria and European Turkey. *Environ. Monit. Assess.*, *157*, 529-537.

Couto, J. A., Frenandez, J., Aboal, J. R. & Carballeira, A. (2004). Active biomonitoring of element uptake with terrestrial mosses: A comparison of bulk and dry deposition. *Sci. Total Environ.*, *324*, 211-222.

Crock, J. G., Beck, K. A., Fey, D. L., Hageman, P. L., Papp, C. S. & Peacock, T. R. (1993). *Element concentrations and baselines for moss, lichen, spruce and surface soils in and near Wrangell–Saint Elias National Park and Preserve, Alaska.* U.S. Geological Survey Open-File Report: Denver, 93-14.

Crock, J. G., Gough, L. P., Mangis, D. R., Curry, K. L., Fey P. L., Hageman, P. L. & Welsch, E. P. (1992). *Element concentrations and trends for moss, lichen, and surface soils in and near Denali National Park and Preserve, Alaska.* U.S. Geological Survey Open-File Report: Denver, 92-323.

Ćujić, M., Deagović, S., Sabovljević, M., Slavković-Beškoski, L., Kilibarda, M., Savivić, J. & Onija, A. Use of mosses as biomonitors of major, minor and trace element deposition around the largest thermal power plant in Serbia. *CLEAN–Soil Air Water*, [Online early access]. DOI: 10.1002/clen.201100656. Published Online: July 3, 2013. http://onlinelibrary.wiley.com.

Dołęgowska, S. & Migaszewski, Z. M. (2013). Anomalous concentrations of rare earth elements in the moss-soil system from south-central Poland. *Environ. Pollut.*, *178*, 33-40.

Dołęgowska, S., Migaszewski, Z. M. & Michalik, A. (2013). *Hylocomium splendens* (Hedw.) B.S.G. and *Pleurozium schreberi* (Brid.) Mitt. as trace element bioindicators: statistical comparison of bioaccumulative properties. *J. Environ. Sci.*, 25, 340-347.

Dołęgowska, S. & Migaszewski, Z. M. (2011). PAH concentrations in the moss species *Hylocomiun splendens* (Hedw.) B.S.G. and *Pleurozium schreberi* (Brid.) Mitt. Form the Kielce area (south-central Poland). *Ecotoxicol. Environ. Saf.*, 74, 1636-1644.

Dragović, S. & Mihailović, N. (2009). Analysis of mosses and topsoils for detecting sources of heavy metal pollution: multivariate and enrichment factor analysis. *Environ. Monit. Assess.*, *157*, 383-390.

Ermakova, E. V., Frontsayeva, M. V., Pavlov, S. S., Povtoreiko, E. A., Steinnes, E. & Cheremisina, Ye. N. (2004). Air pollution studies in Central Russia (Tver and Yaroslav Regions) using the moss biomonitoring technique and neutron activation analysis. *J. Atmos. Chem.*, *49*, 549-561.

Ernst W. H. O. (1995). Sampling of plant material for chemical analysis. *Sci. Total Environ.*, *176*, 15-24.

Fernández, J. A., Aboal, J. R. & Carballeira, A. (2000). Use of native and transplanted mosses as complementary techniques for biomonitoring mercury around an industrial facility. *Sci. Total Environ.*, *256*, 151-161.

Fernández, J. A. (2002). Carballeira, A. Biomonitoring metar deposition in Galicia (NW Spain) with mosses: factors affecting bioconcentration. *Chemosphere*, *46*, 535-542.

Ford, J., Landers, D., Kugler, D., Lasorsa, B., Crecelius, E. & Martinson, J. (1995). Inorganic contamination in arctic Alaskan ecosystems: long range atmospheric transport or local point sources. *Sci. Total Environ.*, *160*, 323-335.

Gadzała-Kopciuch, R., Berecka, B., Bartoszewicz, J. & Buszewski, B. (2004). Some consideration about bioindicators in environmental monitoring. *Pol. J. Environ. Stud.*, *13*, 453-462.

Galsomiès, L., Ayrault, S., Carrot, F., Deschamps, C. & Letrouit-Galinou, M. A. (2003). Interspecies calibration in mosses at regional scale – heavy metal and trace elements results from Ile-de-France. *Atmos. Environ.*, *37*, 241-251.

Gałuszka, A. (2007). Distribution patterns of PAHs and trace elements in mosses *Hylocomium splendens* (Hedw.) B.S.G. and *Pleurozium schreberi* (Brid.) Mitt. from different forest communities: A case study, south-central Poland. *Chemosphere*, *67*, 1415-1422.

Garty, J., Karary, Y. & Harel, J. (1993). The impact of air pollution on the integrity of cell membranes and chlorophyll in the lichen *Romalina dutiaei* (De Not.) Bagl. Transplanted to industrial sites in Israel. Archives of *Environ. Monit. Assess.*, *14*, 77-87.

Gerdol, R., Bragazza, L. & Marchesini, R. (2002). Element concentrations in the forest moss *Hylocomium splendens*: variation associated with altitude, net primary production and soil chemistry. *Environ. Pollut.*, *116*, 129-135.

Gough, L. P. & Crock J. G. (1997). *Distinguishing between natural geologic and anthropogenic trace element sources, Denali National Park and Preserve.* U.S. Geological Survey Professional Paper: Denver, 1574, 57-71.

Grodzińska, K. & Szarek-Łukaszewska, G. (2001). Response of mosses to the heavy metal deposition in Poland – an overview. *Environ. Pollut.*, *114*, 443-451.

Grodzińska, K., Szarek-Łukaszewska, G. & Godzik, B. (1999). Survey of heavy metal deposition in Poland using mosses as indicators. *Sci. Total Environ.*, *229*, 41-51.

Halleraker, J. H., Reimann, C., Caritat, P., Finne, T. E., Kashulina, G., Niskavaara, H. & Bogatyrev, I. (1998). Reliability of moss (*Hylocomium splendens* and *Pleurozium schreberi*) as a bioindicators of atmospheric chemistry in the Barents region: Interspecies and field duplicate variability. *Sci. Total Environ.*, *218*, 123-139.

Harmens, H., Buse, A., Büker, P., Norris, D., Mills, G., Williams, B., Reynold, B., Ashende, T. W., Rühling, Å. & Steinnes, E. (2004). Heavy metal concentrations in European mosses: 2000/2001 survey. *J. Atmos. Chem.*, *49*, 425-436.

Harmens, H., Foan, L., Simon, V. & Mills, G. (2013). Terrestrial mosses as biomonitors of atmospheric POPs pollution: a review. *Environ. Pollut.*, *173*, 245-254.

Harmens, H., Norris, D., Koerber, G. R., Buse, A., Steinnes, E. & Rühling, Å. (2008). Temporal trends (1990-2000) in the concentration of cadmium, lead and mercury in mosses across Europe. *Environ. Pollut.*, *151*, 368-376.

Holoubek, I., Klánová, J., Jarkovský, J., Kubik, V. & Helešic, J. (2007). Trends in background levels of persistent organic pollutants at Kosetice observatory, Czech Republic. *J. Environ. Monitor.*, *9*, 564-571.

Holoubek, I., Korinek, P., Seda, Z., Schneiderová, E., Holoubková, I., Pacl, A., Triska, J., Cudlin P. & Cáslavsky J. (2000). The use of mosses and pine needles to detect persistent organic pollutants at local and regional scales. *Environ. Pollut.*, *109*, 283-292.

Holy, M., Leblond, S., Pesch, P. & Schröder, W. (2009). Assessing spatial patterns of metal bioaccumulation in French mosses by means of an exposure index. *Environ. Sci. Pollut. R.*, *16*, 499-507.

Huttunen, S., Taipale, T., Lappalainen, N. M., Kubin, E., Lakkala, K. & Kaurola, J. (2005). Environmental specimen bank samples of *Pleurozium schreberi* and *Hylocomium splendens* as indicators of the radiation environment at the surface. *Environ. Pollut.*, *133*, 315-326.

Kabata-Pendias, A. & Pendias H. (1999) *Biogeochemistry of trace elements*. PWN: Warszawa.

Köhler, J. & Peichl, L. (1993). *Moose als Bioindikatoren für Schwermetalle.* Bayerisches Staatsministerium für Landesentwicklung und Umweltfragen: München.

Kosior, G., Samecka-Cymerman, A., Kolon, K. & Kempers, A. J. (2010). Bioindication capacity of metal pollution of native and transplanted *Pleurozium schreberi* under various levels of pollution. *Chemosphere, 81*, 321-326.

Leblond, S., Gombert, S., Colin, J. L., Losno, R. & Rausch de Traubenberg, C. (2004). Biological and temporal variations of trace element concentrations in the moss species *Scleropodium purum* (Hedw.) Limpr. *J. Atmos. Chem., 49*, 95-110.

Lehendorff, E. & Schwark, L. (2008). Accumulation histories of major and trace elements on pine needles in the Cologne Conurbation as function of air quality. *Atmos. Environ., 42*, 833-845.

Lodenius, M. (2013). Use of plants for biomonitoring of airborne mercury in contaminated areas. *Environ. Res., 25*, 113-123.

Loppi, S. & Bonini, I. (2000). Lichens and mosses as biomonitors of trace elements in areas with thermal springs and fumarole activity (Mt. Amiata, central Italy). *Chemosphere, 41*, 1333-1336.

Makholm, M. M. & Mladenoff, D. J. (2005). Efficacy of a biomonitoring (moss bag) technique for determining element deposition trends on a mid-range (375 km) scale. *Environ. Monit. Assess., 104*, 1-18.

Mäkinen, A. (1994). Biomonitoring of atmospheric deposition in Finland, Estonia and the Kola Peninsula, based on the chemical analysis of mosses. Ph.D. Dissertation, University of Helsinki, Helsinki.

Mariet, C., Gaudry, A., Ayrault,·S., Moskura, M., Denayer,·F. & Bernard, N. (2011). Heavy metal bioaccumulation by the bryophyte *Scleropodium purum* at three French sites under various influences: rural conditions, traffic, and industry. *Environ. Monit. Assess., 174*, 107-118.

Marinova, S., Yurukova, L., Frontsayeva, M. V., Steinnes, E., Strelkova, L. P., Marinov, A. & Karadzhinova, A. G. (2010). Air pollution studies in Bulgaria using the moss biomonitoring technique. *Eco. Chem. Eng. S., 17*, 37-52.

Markert, B. (2007). Definitions and principles for bioindication and biomonitoring of trace metals in the environment. *J. Trace Elem. Med. Biol., 21*, 77-82.

Markert, B. & Weckert, V. (1989). Use of *Polytrichum formosum* (moss) as a passive biomonitor for heavy metal pollution (cadmium, copper, lead and zinc). *Sci. Total Environ., 86*, 289-294.

Migaszewski, Z. M., Gałuszka, A, Dołęgowska, S., Crock J. G. & Lamothe P. J. (2010). Mercury in mosses *Hylocomium splendens* (Hedw.) B.S.G. and *Pleurozium schreberi* (Brid.) Mitt. from Poland and Alaska: understanding the origin of pollution sources. *Ecotoxicol. Environ. Saf.*, *73*, 1354-1351.

Migaszewski, Z. M., Gałuszka, A., Crock, J. G., Lamothe, P. J. & Dołęgowska, S. (2009). Interspecies and interregional comparison of the chemistry of PAHs and trace elements in mosses *Hylocomium splendens* (Hedw.) B.S.G. and *Pleurozium schreberi* (Brid.) Mitt. from Poland and Alaska. *Atmos. Environ.*, *43*, 1464-1473.

Migaszewski Z. M., Gough L. P. & Gałuszka A. (2001). The Role of Biogeochemistry in Environmental Studies. *Geol. Quart.*, *49(10/2)*, 960-965.

Nieboer, E. & Richardson, D. H. S. (1981). The replacement of the nondescript term 'heavy metals' by a biologically and chemically significant classification of metal ions. *Environ. Pollut.*, *1*, 3-26.

Onianwa, P. C. (2001). Monitoring atmospheric metal pollution: a review of the use of mosses as indicators. *Environ. Monit. Assess.*, *71*, 13-50.

Orliński, R. (2002). Multipoint moss passive samplers assessment of urban airborne polycyclic aromatic hydrocarbons: concentrations profile and distribution along Warsaw main streets. *Chemosphere*, *48*, 181-186.

Osborn, D., Weeks, J. M., Hankard, P. & Dale, L. (2000). *Potential uses of biomonitoring in pollution control-An introductory guide*. Environmental Agency Technical Report: Bristol, UK.

Ötvös, E., Kozák, I. O., Fekete, J., Sharma, V. K. & Tuba, Z. (2004). Atmospheric deposition of polycyclic aromatic hydrocarbons (PAHs) in mosses *(Hypnum cupressiforme)* in Hungary. *Sci. Total Environ.*, *330*, 89-99.

Pesch, P. & Schröder, W. (2006). Mosses as bioindicators for metal accumulation: Statistical aggregation of measurement data to exposure indices. *Ecol. Indic.*, *6*, 137-152.

Poikolainen, J., Kubin, E., Piispanen, J. & Karhu, J. (2004). Atmospheric heavy metal deposition in Finland during 1985-2000 using mosses as biomonitors. *Sci. Total Environ.*, *318*, 171-185.

Poikolainen, J., Piispanen, J., Karhu, J. & Kubin, E. (2009). Long-term changes in nitrogen deposition in Finland (1990-2006) monitored using the moss *Hylocomium splendens*. *Environ. Pollut.*, *157*, 3091-3097.

Polikolainen, J. (2004). Mosses, epiphytic lichens and tree bark as biomonitors for air pollutants – specifically for heavy metals in regional surveys. Oulu University Press: Oulu.

Prasad, M. N. V. (2008). Trace elements as contaminants and nutrients: consequences in ecosystems and human health. Wiley: New York.

Prudêncio, M. I. (2006). Biogeochemistry of trace and major elements in a surface environment (volcanic rock, soil, mosses, lichens) in the S. Miguel Island, Azores, Portugal. *J. Radioanal. Nucl. Chem.*, *271*, 431-437.

Reimann, C., Arnoldussen, A., Boyd, R., Finne, T. E., Nordgulen, Ø., Volden, T. & Englmaier, P. (2006). The influence of a city on element contents of a terrestrial moss (*Hylocomium splendens*). *Sci. Total Environ.*, *369*, 419-432.

Reimann, C., Niskavaara, H., Kashulina, G., Filzmoser, P., Boyd, R., Volden, T., Tomilina, O. & Bogatyrev, I. (2001). Critical remarks on the use of terrestrial moss (*Hylocomium splendens* and *Pleurozium schreberi*) for monitoring of airborne pollution. *Environ. Pollut.*, *113*, 41-57.

Rivera, M., Zechmeister, H., Medina-Ramón, M., Basagańa, X., Foraster, M., Bouso, L., Moreno, T., Solanas, P., Ramos, R., Köllensperger, G., Deltell, A., Vizcaya, D. & Künzli, N. (2011). Monitoring of heavy metal concentrations in home outdoor air using moss bags. *Environ. Pollut.*, *159*, 954-962.

Ross, H. B. (1990). On the use of mosses (*Hylocomium splendens* and *Pleurozium schreberi*) for estimating atmospheric trace metal de position. *Water, Air, Soil Pollut.*, *50*, 63-76.

Ross, S. E., Callaghan, T. V., Ennos, A. R. & Sheffield E. (1998). Mechanics and growth form of the moss *Hylocomium splendens*. *Ann. Bot.*, *82*, 787-793.

Rucandio, M. I., Petit-Dominguez, M. D., Fidalgo-Hijano, C. & Garcia-Gimenez, R. (2011). Biomonitoring of chemical elements in an urban environment using arboreal and bush plant species. *Environ. Sci. Poll. Res.*, *18*, 51-63.

Rühling, A. (1989). Survey if the heavy-metal deposition in Europe using bryophytes as bioindocators. Proposal for an international programme. IVL/Dept. of Plant Ecology. University of Lund.

Rühling, Å. & Tyler, G. (1970). Sorption and retention of heavy metals in the woodland moss *Hylocomium splendens* (Hedw.). *Oikos*, *21*, 922-997.

Rühling, Å. & Tyler, G. (1973). Heavy metal deposition in Scandinavia. *Water, Air, Soil Pollut.*, *2*, 445-455.

Rühling, Å. & Tyler, G. (2004). Changes in the atmospheric deposition of minor and rare elements between 1975-2000 in south Sweden, as measured by moss analysis. *Environ. Pollut.*, *131*, 417-423.

Sabovljević, M., Vukojević, V., Sabovljević, A. & Vujičić, M. (2009). Deposition of heavy metals (Pb, Sr and Zn) in the country of Obrenovac (Serbia) using mosses as bioindicators. *J. Ecol. Environ. Sci.*, *1*, 147-155.

Saitanis, C. J., Frontsayeva, M. V., Steinnes, E., Palmer, M. W., Ostrovnaya, T. M. & Gundorina, S. F. (2013). Spatiotemporal distribution of airborne elements monitored with the moss bags technique in the Greater Thriasion Plain, Attica Greece. *Environ. Monit. Assess.*, *185*, 955-968.

Šakalys, J., Kvietkus, K., Sucharová, J., Suchara, L. & Valiulis, D. (2009). Changes in total concentrations and assessed background concentrations of heavy metals in moss in Lithuania and the Czech Republic between 1995 and 2000. *Chemosphere*, *76*, 91-97.

Salo, H., Bućko, M. S., Vaahtovuo, E., Mäkinen, J. & Pesonen, L. J. (2012). Biomonitoring of air pollution in SW Finland by magnetic and chemical measurements of moss bags and lichens. *J. Geochem. Explor.*, *115*, 69-81.

Samecka-Cymerman, A., Kosior, G. & Kempers, A. J. (2006). Comparison of the moss *Pleurozium schreberi* with needles and bark of *Pinus sylvestris* as biomonitors of pollution by industry in Stalowa Wola (southeast Poland). *Ecotoxicol. Environ. Saf.*, *65*, 108-117.

Schröder, W. & Pesch, R. (2010). Long-term monitoring of metal accumulation in forests measured by use of the moss technique. *Eur. J. Forest Res.*, *129*, 475-488.

Schröder, W., Pesch, R., Englert, C., Harmens, H., Suchara, I., Zechmeister, H. G., Thöni, L., Maňkovská, B., Jeran, Z., Grodzińska, K. & Alber, R. (2008). Metal accumulation in mosses across national boundaries: Uncovering and ranking causes of spatial variation *Environ. Pollut.*, *151*, 377-388.

Siewers, U. & Herpin, U. (1998). *Schwermetalleinträge in Deutschland. Moos-Monitoring 1995/96.* Geologisches Jahrbuch, Sonderhefte, Reihe D, Heft SD 2, Schweizerbarth'sche Verlagsbuchhandlung: Stuttgart, Germany.

Siontorou, C. G. & Batzias, F. A. (2011). Error identification/ propagation/remediation in biomonitoring survey – A knowledge-based approach towards standardization via fault tree analysis. *Ecol. Indic.*, *11*, 564-581.

Sloof, J. E. (1995). Lichens as quantitative biomonitors for atmospheric trace-element deposition, using transplants. *Atmos. Environ.*, *29*, 11-20.

Smagunova, A. N., Karpukova, O. M., Belykh, L. I., Malykh, Y. M., Aprelkova, N. F. & Kozlov, V. A. (2004). Sampling errors in monitoring the pollution of environmental samples. *Fresenius J. Anal. Chem., 59*, 1146-1152.

State, G., Popescu, I. V., Radulescu, C., Macris, C., Stihi, C., Gheboianu, A., Dulama, I. & Nitescu, O. (2012). Comparative studies of metal air pollution by atomic spectrometry techniques and biomonitoring with moss and lichens. *Bull. Environ. Contam. Toxicol., 89*, 580-586.

Steinnes, E. (1995). A critical evaluation of the use of naturally growing moss to monitor the deposition of atmospheric metals. *Sci. Total Environ., 160/161*, 243-249.

Suchara, I. & Sucharová, J. (1998). Atmospheric deposition levels of chosen elements in the Czech Republic determined in the framework of the International Bryomonitoring Program 1995. *Sci. Total Environ., 223*, 37-52.

Sucharová, J & Suchara, I. (2004). Current multi-element distribution in forest epigeic moss in the Czech Republic-a survey of Czech national biomonitorng programme 2000. *Chemosphere, 57*, 1389-1398.

Szczepaniak, K. & Biziuk, M. (2003). Aspects of the biomonitoring studies using mosses and lichens as indicators of metal pollution. *Environ. Res., 93*, 221-230.

Tabors, G., Brumelis, G., Lapina, L., Pospelova, G. & Nikodemus, O. (2004). Changes in element concentrations in moss segments after cross-transplanting between a polluted and non-polluted site. *J. Atmos. Chem., 49*, 191-197.

Thöni, L., Schnyder, N. & Krieg, F. (1996). Comparison of metal concentrations in three species of mosses and metal freights in bulk precipitation. *Fresenius J. Anal. Chem., 354*, 703-708.

Thöni, L., Yurukowa, L., Bergamini, A., Ilyin, I. & Mtthaei, D. (2011). Temporal trends and spatial patterns of heavy metal concentrations in mosses in Bulgaria and Switzerland: 1990-2005. *Atmos. Environ., 45*, 1899-1912.

Tyler, G. (2004). Rare earth elements in soil and plant systems–A review. *Plant Soil, 267*, 191-206.

Ugulu, I., Dogan, Y., Baslar, S. & Varol, O. (2012). Biomonitoring of trace element accumulation in plants growing at Murat Mountain. *Int. J. Environ. Sci. Technol., 9*, 527-534.

Uğur, A., Özden, B., Saç, M. M., Yener, G., Altmbaş, Ü., Kurucu, Y. & Bolca, M. (2004). Lichens and mosses for correlation between trace elements and ^{210}Po in the areas near coal-fired power plant at Yatağan, Turkey. *J. Radioanal. Nucl. Chem., 259*, 87-92.

Wolterbeek, B. (2002). Biomonitoring of trace element air pollution: principles, possibilities and perspectives. *Environ. Pollut.*,*120*, 11-21.

Wolterbeek, H. T., Bode, P. & Verburg T. G. (1996). Assessing the quality of biomonitoring via signal-to-noise ratio analysis. *Sci. Total Environ.*, *180*, 107-116.

Wolterbeek, H. T., Kuik, P., Verburg, T. G., Herpin, U., Markert, B. & Thöni, L. (1995). Moss interspecies comparisons in trace element concentrations. *Environ. Monit. Assess.*, *35*, 263-286.

Yu, M. H., Tsunoda, H. & Tsunoda, M. (2005). *Environmental Toxicology: Biological and Health Effects of Pollutants*, 3rd ed.; CRC Press Inc.: Florida.

Zechmeister, H. G. (1995). Correlation between altitude an heavy metal deposition in Alps. *Environ. Pollut.*, *89*, 73-80.

Zechmeister, H. G. (1998). Annual growth of four pleurocarpous moss species and their applicability for biomonitoring heavy metals. *Environ. Monit. Assess.*, *52*, 441-451.

Zechmeister, H. G., Hohenwallner, D., Riss, A. & Hanus-Illnar, A. (2003). Variations in heavy metal concentrations in the moss species *Abietinella abietina* (Hedw.) Fleisch. according to sampling time, within site variability and increase in biomass. *Sci. Total Environ.*, *301*, 55-65.

Zhou, Q., Zhang, J., Fu, J., Shi, J. & Jiang, G. (2008). Biomonitoring: an appealing tool for assessment of metal pollution in the aquatic ecosystem. *Anal. Chim. Acta*, *606*, 135-160.

In: Moss
Editor: Jorma Mohamed
ISBN: 978-1-63117-396-7

Chapter 3

Moss Bioreactors: Novel Biotechnological Tools for the Production of Pharmaceutical Molecules

***Patrícia B. Pelegrini*[1] *and Maria Fátima Grossi de Sa*[2,3]**
[1]BioLife Brasil Ltda, Formosa-GO, Brazil
[2]Research National Centre for Genetic Resources and Biotechnology, Brasilia, DF, Brazil
[3]Catholic University of Brasilia, Brasilia, DF, Brazil

Abstract

The use of moss bioreactors for the production of proteins and peptides is a recently incorporated heterologous system that was developed as an alternative for the expression of recombinant molecules that were not well obtained in mammals, insect and/or yeast systems. Moss is classified into plant kingdom, but it contains a very small genome of 511 Mbps, which facilitates its manipulation and gene transformation. Moreover, it presents less risk of contaminations with human and animal pathogens, when compared with expression systems using bacterial and mammalian cells. Nowadays, the moss species most used as a bioreactor is *Physcomitrella patens*. It was earlier used for studies on plant evolution, development and physiology. Since the last 30 years, *P. patens* started being evaluated as a new option for the production of recombinant proteins. The advantages of moss over other higher plant species is the enhanced rates of homologous recombination, giving straight-forward reverse genetics protocols by gene targeting. Furthermore, it lacks a bias for codon usage, making it less complicated to adapt foreign genes in its genome for protein expression. The cultivation of *P. patens* is also relatively simple and of low cost, when compared to higher plant species. It can grow under a wide range of pH and temperature, using light and carbon dioxide as its only energy and carbon sources. Until now, most proteins expressed in moss systems are from the human genome, with the purpose of producing biopharmaceutical molecules. Hence, it has been described the use of moss bioreactors for the production of monoclonal antibodies, vaccines-to-be molecules and other human proteins and peptides. Therefore, this report aims to describe the most recent advances on the use of moss bioreactor for the expression of proteins targeted as future biopharmaceutical molecules.

Keywords: Moss, bioreactor, biopharmaceutical, proteins, peptides

INTRODUCTION

The need for the production of therapeutic proteins in large scale started during the Second World War, when it was observed an increase search for blood-derived proteins [1]. The importance of biopharmaceuticals is incontestable, but many proteins with therapeutic activity were obtained through purification of animal sources, leading to potential allergenic risks to human health.

Hence, the first biopharmaceutical molecule produced was insulin, which was obtained from the purification of animal pancreas [2]. Nowadays, biopharmaceuticals are produced through techniques involving biotechnological processes, resulting in more efficient, safe and stable proteins. Further studies on expression vectors and promoters for the increase of protein production allowed the entrance of a new field into the optimization of protein expression in heterologous systems [3-5].

Since then, several other organisms, including yeast, mammal cells and plants were studied in order to better understand their specific genomic properties and cell machinery for the best performance on protein production in heterologous systems.

In this way, bacteria species of *Escherichia coli* were the first to be used as a host cell on protein expression. Due to the wide knowledge on their molecular biology and genetic features, *E. coli* cells were also the first organism to be utilized in bioreactors for scaled up production of foreigner proteins [4-9]. However, the application of this system for certain eukaryotic proteins showed to be a problem, as prokaryotic organisms do not present the complex tools needed for post-translational processes, neither a secretion way that allow the transportation of recombinant proteins to the extracellular medium.

Therefore, in order to obtain proteins that require such complex post-translational processes, other organisms were chosen and characterized for protein expression. Thus, yeast, mammal cells, insect cells and later, plants, brought up alternatives for the production of eukaryotic recombinant proteins in large scale [8, 12-15]. These systems allowed that production of human proteins with Disulfide Bridge, glycosylation and phosphorylation pattern could be expressed in foreigner hosts.

Hence, during the last three decades, the production of proteins and peptides in heterologous systems has been intensively studied, generating great advances in the area. Nevertheless, in terms of different number of possible hosts for protein expression, there was not a significant increase observed since the 1980´s.

The biggest change was the enhancement of the protein concentration produced in the cell host, which increased due to the introduction of bioreactors into the heterologous systems [16, 17]. Other change detected was the modification of culture medium and the application of different feeding strategies on the cells/tissues, which permitted a protein expression in a constitutive way.

Recent studies highlighted the use of bioreactors as the best cost-efficiency process for the production of high concentrations of recombinant proteins, which can be applied into different cell types and organisms [17-21]. Bioreactors are highly indicated for the production of biopharmaceutical products, as they are important not only to the industry but also for the public health system [15, 18, 20, 22-26]. In this way, plants occupy the first position of preference, especially for the production of human proteins for therapeutic use [27-29].

More recently, the use of moss species for the expression of biopharmaceuticals showed to be an efficient and cost-effective method for large-scale production of human proteins [18, 23, 30].

Therefore, this chapter describes a review of plant bioreactors for the production of biopharmaceutical proteins, with emphasis in the moss system.

Moss Cultivation and Application into Bioreactors for Protein Expression

The cultivation of *P. patens* is a relatively simple and low-cost procedure. *In vitro* cultivation is the best option for mosses, and involves haploid pore germination that produces protonema cells, which proliferate through apical cellular division. Hence, protonema division will stimulate the development of buds and sprouts that will become the gametophyte and, later, the adult plant [18]. The formation of sexual organs in moss can lead to self-fecundation and the development of encapsulate spores at the top of the gametophyte. These spores will, then, be released, completing the moss life cycle [18, 31].

Moss culture medium is essentially made of water and inorganic salts and, although many organic and inorganic compounds were tested for the optimization of its growth, there is no need of adding hormones or other complex additives to the medium [32]. Light and carbonic gas are the only energy sources necessary for moss growth, and they can grow under a wide range of pH and temperature [32, 33].

The technique of protein expression in plants, including mosses, is highly extensive and depends of several variables, such as: (i) the choice of a plant cell line; (ii) features of the protein to be expressed; (iii) details on the plasmid construction (choice of best vector, promoter, signal peptide, etc.); (iv) Choice of methodology for transformation and regeneration; (v) methods for purification and characterization of the recombinant protein.

Moss maintenance by cultivation of the diploid forms of the organism in solid medium is the most used (Figure 1A). However, protein expression is performed using the haploid protonema of the moss in liquid medium. In bioreactors, it is recommended to use liquid medium culture for moss systems. Therefore, protonema of *P. patens* are first incubated in solid medium, such as $PpNH_4$ culture medium. Later, the adult plant is selected and the protonema are separated for application into bioreactors containing liquid medium culture (Figure 1B). Usually, an initial moss dry weight of 30mg per liter of medium is used to initiate the protein expression [34].

Some factors are essential for the optimization of protein expression processes in moss, especially when the final product will be applied in the biopharmaceutical industry. Among them, there is the pH of the culture medium, which should not vary during protein expression. Thus, the pH of 5.8 demonstrated to be the best for moss cells growth and protein production. In some cases, pH 4.5 also showed to be efficient for the expression of certain human proteins in *P. patens* [35] (Table 1). Furthermore, the expression of the recombinant protein occurs usually after 4 days of moss inoculation into the culture medium. Therefore, moss harvesting for protein extraction is done up to 12 days after moss inoculation [33-34]. However, cell growth must be constantly checked through measurement of moss dry weight (50mL/liter of culture medium), which can be done by incubation of cells at 105° C for 2 hours.

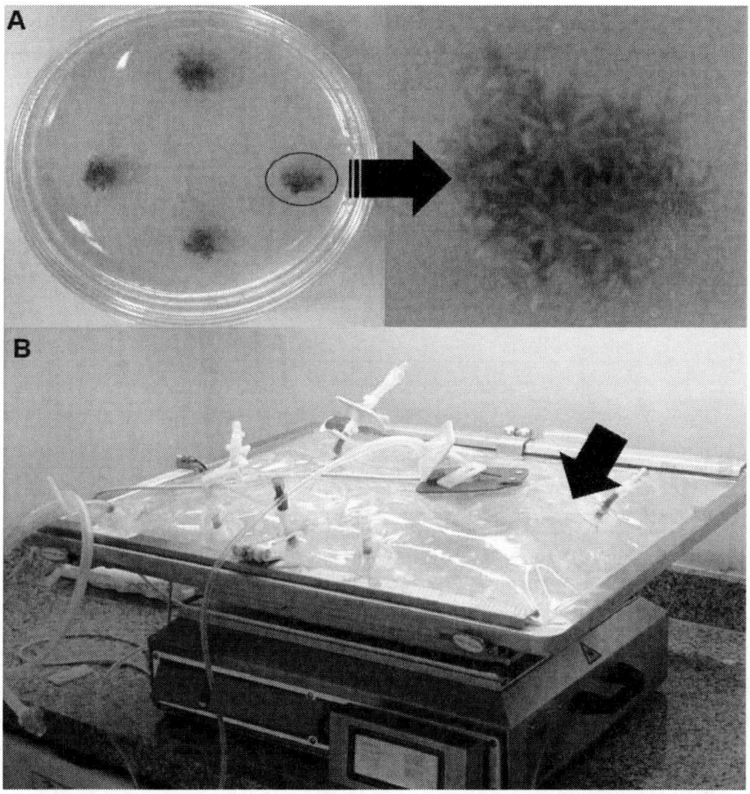

Figure 1. A) Moss cultivation into solid medium $PpNH_4$.Arrow indicates amplified view of the adult form of *P. patens* during growth in solid medium plate. B) Moss cultivation into liquid medium Knop for protein expression using Wave bioreactor (GE).Arrow indicates the plastic bag (maximum capacity of 10L) containing 8L of culture medium and moss cells for protein expression at an initial mass rate of 50mg/L. Light/dark periods are controlled by a LED system installed over the bioreactor.

Light conditions are extremely important for moss growth, and are very similar to what is needed for higher plants in greenhouses: 16 hours of light and 8 hours of dark. Moreover, although ideal temperature conditions for moss grow is 25° C, earlier studies showed that, at 22° C, lines of *P. patens* can still stimulate protein expression [36]. In general, the insertion of oxygen and carbon dioxide is not a rule, but an addition of 1ppm of atmospheric air can stimulate cell mass growth and it is sufficient to maintain the moss under optimal conditions of survivance. Nevertheless, for cylindrical bioreactor, it is recommended to add 0.3vvm of oxygen and 2% of carbon dioxide into moss culture media under agitation of 100rpm [34] (Table 1).

Table 1. Growth conditions for protein expression using moss cell system

Conditions	**Value range**
Temperature	22-25° C
Light/dark	16h/8h
pH	4.5-5.8
pO2	0.3vvm
pCO2	2%

Application of Moss Bioreactors into Biopharmaceutical Industry

During the last 10 years, the moss *P. patens* was introduced as a novel tool for the expression of biopharmaceutical proteins due to its facility of cell culture and adaptation into bioreactors [18, 23, 30]. Similar to what happened with higher plants, several techniques on genetic engineering have been applied into mosses in order to remove undesired glycosylation processes. Moreover, the addition of N-glycan molecules analogous to the ones observed in humans were done in some moss lines, so the system be able to perform post-translational activities related to what is observed in human cells [37, 38].

The glycosylation system in moss is analogous to the process that occurs in higher plants [39]. The main central structure of oligosaccharides in species of *P. patens* is formed by N-acetilglycosamine (NAcGlc) and mannose, with two galactose molecules linked to the extremity of two NAcGlc residues. This base structure is equal in both plant and mammal species. The difference is that, in mosses and plants, there is a β-1,2 interaction between one mannose and one xylose molecule, two α-1,4 interactions between a fucose and a NAcGlc molecule, and two β-1,3 between a NAcGlc and a galactose molecule [39]. In mammals, the difference lays on the absence of xylose molecules and the addition of β-1,6 NAcGLc-fucose and β-1,4 NAcGlc-galactose linkages [40]. These differences could be sufficient to cause allergenic responses in humans by proteins containing the glycosylation pattern of plants and mosses [41, 42] (Figure 2).

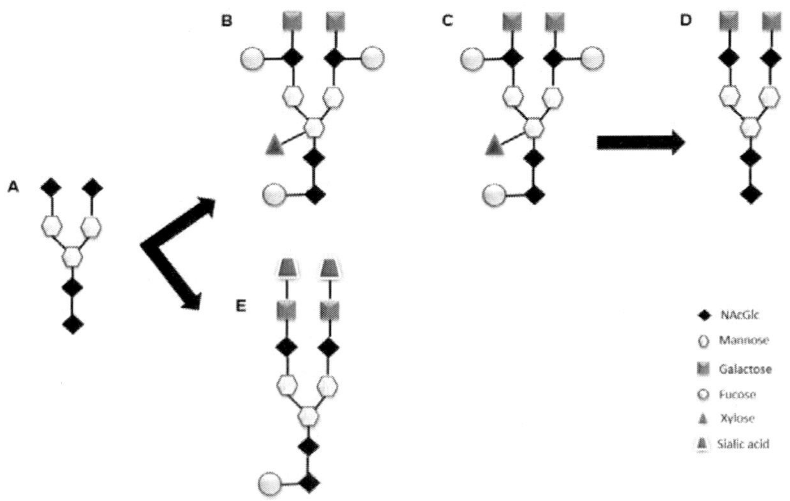

Figure 2. Different forms of glycosylation processing of proteins. A) N-glycan structure of glycosylation profile similar to all organisms. B) N-glycan structure in plants. C) N-glycan structure in moss. D) N-glycan structure of mutant mosses. E) N-glycan structure in mammals.

Earlier studies demonstrated the use of genetic engineering and the optimization of culture conditions allowed successful advances on the yield of products developed into moss systems. Therefore, genetically modified mosses with absence of xylose and fucose molecules were developed in order to produce human proteins for therapeutic use that did not cause allergenic responses [37-39, 43]. Two examples of human proteins already produced in these systems are the human factor H and VEGF [37, 44-47].

Lately, due to its number of advantages, the moss *P. patens* is the most used species for bioreactor systems and protein expression. Hence, one of the benefits of this organism compared to other eukaryotes is the need of simple culture medium for its growth, which contains only inorganic minerals and is absent of hormones, complex carbohydrate and amino acid sources [18]. Another advantage of using *P. patens* as a bioreactor is the high nuclear DNA recombination rate, allowing the knockout of specific genes and the development of new cell lines, as well as the high adaptability for protein production, as codon usage is not necessary for gene [23, 48-52]. Moreover, as it is a model organism for physiological studies, the knowledge of *P. patens* growth and development is already well established, which permits better evaluation of its behavior in the environment [53]. The fact that the major development stage of mosses is during a haploid gametophyte form, glycosylation processes are easier to be controlled, when compared to other cell cultures [49, 54], as described earlier.

Hence, the use of *P. patens* cultures as bioreactors initiated due to their ability of performing post-translational processes, including glycosylation and disulfide bridge formation. Later, it showed to be an excellent tool on the production of biopharmaceuticals and other biotechnological products [55]. During the period of less than ten years, moss bioreactors arose as a novel alternative for the production of recombinant proteins in large-scale, especially for the development of biopharmaceutical products. As it has a small genome of 511 Mbps, it is an easier system for DNA manipulation than other higher plants. In addition, the risk of contamination by human pathogens is reduced, when compared to bacteria and mammal cell cultures.

Furthermore, moss cultivation in glass bioreactors are already being established and demonstrated to be more adaptable to large-scale protein expression technologies [56]. Although it is a recent technology, when compared to other protein expression systems, such as bacterial, yeast and mammal cells, the moss *P. patens* has already been used for the production of several proteins of therapeutic application. Among them, there is the human H factor, utilized on the treatment of many diseases, including kidney and retinal disorders, hemolytic uremic syndrome (AUS), age-related macular degeneration (ARMD) and membranoproliferative glomerulonephritis II (MPGN II) [47]. Similarly, the glycosylated peptide erythropoietin (EPO), which represents a key role on the regulation of erythrocyte maturation, was produced into the mutant line of *P. patens* Delta-fuc-t Delta-xyl-t [46]. Moreover, it was shown that suspension cells of the moss *P. patens* were an excellent host for the expression of the recombinant endothelial growth factor 121 (rhVEGF), important for the treatment of lesions obtained by diabetic patients [35]. IgG monoclonal antibodies were also produced into moss systems, such as the IGN311 antibody, specific for tumoral antigens [45] (Table 2).

Table 2. Proteins produced using moss bioreactors

Protein	Use	Company/University	Reference
Endotelial growth fator 121 (rhVEGF)	Lesions in diabetic patients	Greenovation Biotech GmbH (Germany)	[35]
Monoclonal antibody IGN311	Anti-tumoral	Igeneon AG (Austria)	[45]
Human H factor	Kidney and retina severe disorders	University of Freiburg (Germany)	[47]
Recombinant growth factor of human keratinocytes (FGF7/KGF)	Mucosite	Greenovation Biotech GmbH (Germany)	[57]
Alpha-galactosidase	Fabry disease	Greenovation Biotech GmbH (Germany)	[58]
Glycocerebrosidase	Gaucher disease	Greenovation Biotech GmbH (Germany)	[58]
Erythropoietin	Regulation of erythrocyte maturation	Greenovation Biotech GmbH (Germany)	[46]

Nowadays, the German company Greenovation Biotechn GmbH (Heibronn) is the pioneer on using moss bioreactor for large-scale production of different human proteins [57]. Recently, they launched into the market the medicament Biokine^{TM} FGF7, composed by the recombinant human keratinocyte growth factor (FGF7/KGF), used on the treatment of mucosite. Besides this molecule, the company produce two other recombinant proteins, which are in pre-clinical trial stage: alpha-galactosidase, for the treatment of Fabry disease, and glucocerebrosidase, for the treatment of Gaucher disease [58] (Table 2).

Conclusion

The use of different technologies based on plant protein production depends highly on the features of the final product. In this way, it is indispensable that the chosen platforms also be connected to a scaling process for industrial production and commercialization. Moreover, the expression system must be compatible with the structural and activity demands of the protein of interest and, therefore, post-translational processes must be considered. Furthermore, the production yield must be investigated, as well as the therapeutic application the recombinant protein will be used for.

In the last ten years, different companies focused on the production of biopharmaceutical products are financing less bacterial, yeast and mammal cell systems and are investing more in the application of plant systems. Nowadays, the leading technologies have crossed the barriers required for optimal protein expression in plants and already started launching into the market plant-based recombinant products. The newly introduced moss technology coupled the facilities of prokaryote systems with the advantages of complex expression eukaryote ones for the production of many human proteins for therapeutic use.

When evaluating the benefits of protein expression systems, companies see the use of plant bioreactors as a way to decrease the costs of molecule production for therapeutic use, which will also affect on the final price of the medicaments. The first protein-based biopharmaceutical produced into mammal cells had a high maintenance cost and low cell mass production, leading to a low protein yield. These features contributed for the high price of these biopharmaceutical products. Nevertheless, after the introduction of plant bioreactors, including moss cells, for the large-scale expression of human proteins, a significant decreased on the price of medicaments is expected to be seen for the next few years. Moreover, with the increase demand of biopharmaceutical companies for the development of new products and the continuous advance of technologies, protein expression process is no longer a challenge but a facilitator for the production of molecules for several applications.

REFERENCES

[1] Walls, D.; McGrath, R. and Loughran, S.T. (2011). A digest of protein purification. *Protein Chromatography: Methods and Protocols, Methods in Molecular Biology*, 681, 3-23.

[2] Ferro, E. S. (2010). Biotecnologia Translacional: Hemopressina e outros peptídeos intracelulares. *Estudos Avançados,* 24, 109-121.

[3] Ho, Y.-S.; Wulff, D. L. and Rosenberg, M. (1983) Bacteriophage lambda protein ceII binds promoters on the opposite face of the DNA helix from RNA polymerase. *Nature,* 304. 703-708.

[4] Young, J. F.; Desselberger, U.; Palese, P.; Fergunson, B.; Shatzman, A. R.; Rosenberg, M. (1983) Efficient expression on influenza virus NS1 nonstructural proteins in *Escherichia coli. Proceedings of National Academy of Sciences,* 80, 6105-6109.

[5] Cheng, S. C.; Kim, R.; King, K.; Kim, S. H. and Modrich, P. J. (1984) Positive-selection cloning vehicle useful for overproduction of hybrid proteins. *Biological Chemistry*, 259, 11571-11575.

[6] Shatzman, A. R. and Rosenberg, M. (1986) Efficient expression of heterologous genes in *Escherichia coli*: the pAS vector system and its applications. *Annals of the New York Academy of Sciences,* 487, 233-248.

[7] Shatzman, A. R. and Rosenberg, M. (1987) Expression, identification and characterization of recombinant gene products in *Escherichia coli. Methods in Enzymology*, 152, 661-673.

[8] Demain, A. L. and Vaishnav, P. (2009) Production of recombinant proteins by microbes and higher organisms. *Biotechnology Advances,* 27, 297-306.

[9] Torpe, K. (2006) Overview of bacterial expression systems for heterologous protein production: from molecular and biochemical fundamentals to commercial systems. *Applied Microbiology and Biotechnology,* 72, 211-222.

[10] Chevallier, M. R. and Aigle, M. (1979) Qualitative detection of penicillinase produced by yeast strains carrying chimeric yeast-coli plasmids. *FEBS Letters,* 108, 179-180.

[11] Chevallier, M. R.; Bloch, J. C. and Lacroute, F. (1980) Transcriptional and translational expression of a chimeric bacterial-yeast plasmid in yeasts. *Gene,* 11, 11-19.

[12] Cohen, J. D.; Eccleshall, T. R.; Needleman, R. B.; Federoff, H.; Buchferer, B. A. and Marmur, J. (1980) Functional expression in yeast

of the *Escherichia coli* plasmid gene coding for chloramphenicol acetyltransferase. *Proceedings of National Academy of Sciences USA,* 77, 1078-1082.

[13] Dickson, R. C. (1980) Expression of a foreign eukaryotic gene in *Saccharomyces cerevisiae:* beta-galactosidase from *Kluyveromyces lactis*. *Gene,* 10, 347-356.

[14] Gunge, N. (1983) Yeast DNA plasmids. *Annual Review of Microbiology,* 37, 253-276.

[15] Viana, A. A.; Pelegrini, P. B.; Grossi-de-Sá, M. F. (2012) Plant biofarming: novel insights for peptide expression. In heterologous systems. *Biopolymers,* 98, 416-427.

[16] Wurm, F. M. (2004) Production of recombinant protein therapeutics in cultivated mammalian cells. *Nature Biotechnology,* 22, 1393-1398.

[17] Hacker, D. L.; de Jesus, M. and Wurm, F. M. (2009) 25 years of recombinant proteins from reactor-grown cells – where do we go from here? *Biotechnology Advances,* 27, 1023-1027.

[18] Decker, E. L. and Reski, R. (2008) Current achievements in the production of complex biopharmaceuticals with moss bioreactors. *Bioprocess and Biosystems Engineering,* 31, 3-9.

[19] Potvin, G. and Zhang, Z. (2010) Strategies for high-level recombinant protein expression in transgenic microalgae: a review. *Biotechnology Advances,* 28, 910-918.

[20] Avesani, L.; Bortesi, L.; Santi, L.; Falorni, A.; Pezzoti, M. (2010) Plant-made pharmaceuticals for the prevention and treatment of autoimmune diseases: where are we? *Expert Review in Vaccines,* 9, 957-969.

[21] Zhou, Y.; Chen, H.; Li, X.; Wang, Y.; Chen, K.; Zhang, S.; Meng, X.; Lee, E. Y. and Lee, M. Y. (2011) Production of recombinant human DNA polymerase delta in a *Bomby mori* bioreactor. *PLoS One,* 6, e22224.

[22] Fisher, R.; Stoger, E.; Schillberg, S.; Christou, P. and Twyman, R. M. (2004) Plant-based production of biopharmaceuticals. *Current Opinion in Plant Biology,* 7, 152-158.

[23] Decker, E. L. and Reski, R. (2007) Moss bioreactors producing improved biopharmaceuticals. *Current Opinion in Biotechnology*, 18, 393-398.

[24] Ahmad, A.; Pereira, E. O.; Conley, A. J.; Richman, A. S. and Menassa, R. (2010) Green biofactories: recombinant protein production in plants. *Recent Plant Biotechnology,* 4, 242-259.

[25] Huang, T-K. and McDonald, K. A. (2012) Bioreactor systems for in vitro production of foreign proteins using plant cell cultures. *Biotechnology Advances,* 30, 398-409.
[26] Shukla, A. A. and Gottschalk, U. (2013) Single-use disposable technologies for biopharmaceutical manufacturing. *Trends in Biotechnology,* 31, 147-154.
[27] Hellwig, S.; Drossard, J.; Twyman, R. M. and Fischer, R. (2004) Plant cell cultures for the production of recombinant proteins. *Nature Biotechnology,* 11, 1415-1422.
[28] Yusibov, V.; Streatfield, S. J. and Kushnir, N. (2011) Clinical development of plant-produced recombinant pharmaceuticals. *Human Vaccines,* 7, 313-321.
[29] Huang, T.K. and McDonald, K. A. (2012) Bioreactor systems for in vitro production of foreign proteins using plant cell cultures. *Biotechnology Advances,* 30, 398-409.
[30] Decker, E. L. and Reski, R. (2012) Glycoprotein production in moss bioreactors. *Plant Cell Reproduction,* 31, 453-460.
[31] Cove, D.; Benzanilla, M.; Harries, P. and Quatrano, R. (2006) Mosses as model systems for the study of metabolism and development. *Annual Review in Plant Biology,* 57, 497-520.
[32] Egener, T.; Granado, J.; Guitton, M. C.; Hohe, A.; Holtorf, H.; Lucht, J. M.; Rensing, S. A.; Schlink, K. Schulte, J.; Schween, G.; Zimmermann, S.; Duwenig, E.; Rak, B. and Reski, R. (2002) High frequency of phenotypic deviations in *Physcomitrella patens* plants transformed with a gene-disruption library. *BMC Plant Biology,* 2, 6.
[33] Reski, R. and Abel, W. O. (1985) Induction of budding on chloronemata and caulonemata of the moss *Physcomitrella patens* using isopentenyladenine. *Planta,* 165, 354-358.
[34] Hohe, A. Reski, R. (2005) Control of growth and differentiation of bioreactor cultures of *Physcomitrella* by environmental parameters. *Plant Cell,* 81, 307-311.
[35] Jost, W.; Link, S.; Horstmann, V.; Decker, E.; Reski, R. and Gorr, G. (2004) Isolation and characterization of three moss derived betatubulin promoters suitable for recombinant expression. *Current Genetics,* 47, 111-120.
[36] Lucumi, A. and Posten, C. (2006) Establishment of long-term perfusion cultures of recombinant moss in a pilot tubular photobioreactor. *Process in Biochemistry,* 41, 2180-2187.

[37] Koprivova, A.; Stemmer, C.; Altmann, F.; Hoffmann, A.; Kopriva, S.; Gorr, G.; Reski, R. and Decker, E. L. (2004) Targeted knockouts of *Physcomitrella* lacking plant-specific immunogenic N-glycans. *Plant Biotechnology,* 2, 517-523.

[38] Huelther, C. M.; Lienhart, O.; Baur, A.; Stemmer, C.; Gorr, G.; Reski, R. and Decker, E. L. (2005) Glyco-engineering of moss lacking plant specific sugar residues. *Plant Biology,* 7, 292-299.

[39] Koprivova, A.; Altmann, F.; Gorr, G.; Kopriva, S.; Reski, R. and Decker, E. L. (2003) N-glycosylation in the moss *Physcomitrella patens* is organized similarly to higher plants. *Plant Biology,* 5, 582-591.

[40] Lerouge, P.; Cabanes-Macheteau, M.; Rayon, C.; Fischette-Laine, A. C.; Gomord, V. and Faye, L. (1998) N-glycoprotein biosynthesis in plants: recent developments and future trends. *Plant Molecular Biology,* 38, 31-48.

[41] Bencurova, M.; Hemmer, W.; Focke-Tejkl, M.; Wilson, I. B. and Altmann, F. (2004) Specificity of IgG and IgE antibodies against plant and insect glycoprotein glycans determined with artificial glycoforms of human transferrin. *Glycobiology,* 14, 457-466.

[42] Jin, C.; Altmann, F.; Strasser, R.; Mach, L.; Schahs, M.; Kunert, R.; Rademacher, T.; Glossl, J. and Steinkellner, H. (2008) A plant-derived human monoclonal antibody induces an anti-carbohydrate immune response in rabbits. *Glycobiology,* 18, 235-241.

[43] Gomord, V.; Fitchette, A. C.; Menu-Bouaouiche, L.; Saint-Jore-Dupas, C.; Plasson, C.; Michaud, D.; Faye, L. (2010) Plant-specific glycosylation patterns in the context of therapeutic protein production. *Plant Biotechnology Journal,* 8, 564-587.

[44] Gorr, G. and Jost, W. (2005) Glycosylation design in transgenic moss for better product efficacy. *Bioprocess Journal,* 4, 26-30.

[45] Schuester, M.; Jost, W.; Mudder, G. C.; Wiederkum, S.; Schwager, C.; Janzek, E., Altmann, F.; Stadlmann, J.; Stemmer, C. and Gorr, G. (2007) In vivo glycol-engineered antibody with improved lytic potential produced by an innovative non-mammalian expression system. *Biotechnology Journal,* 2, 700-708.

[46] Weise, A.; Altmann, F.; Rodriguez-Franco, M.; Sjoberg, E. R.; Bäumer, W.; Launhardt, H.; Kietzmann, M. and Gorr, G. (2007) High-level expression of secreted complex glycosylated recombinant human erythropoietin in the *Physcomitrella* delta-fuc-t Delta-xyl-t mutant. *Plant Biotechnology Journal,* 5, 389-401.

[47] Buttner-Mainik, A.; Parsons, J.; Jerome, H.; Hartmann, A.; Lamer, S.; Schaaf, A.; Schlosser, A.; Zipfel, P. F.; Reski, R. and Decker, E. L. (2011) Production of biologically active recombinant human factor H in *Physcomitrella. Plant Biotechnology Journal,* 9, 373-383.

[48] Schaefer, D. G. (2001) Gene targeting in *Physcomitrella patens. Current Opinion in Plant Biology,* 4, 143-150.

[49] Lorenz, S.; Tintelnot, S.; Reski, R. and Decker, E. L. (2003) Cyclin D-knockout uncouples developmental progression from sugar availability. *Plant Molecular Biology,* 53, 227-236.

[50] Terrier, B.; Courtois, D.; Henault, N.; Cuvier, A.; Bastin, M.; Aknin, A.; Dubreuil, J. and Petiard, V. (2007) Two-phase flow in vertical tubes. *Biotechnology and Bioengineering,* 96, 914-923.

[51] Koprivova, A.; Meyer, A. J.; Schween, G.; Herschbach, C.; Reski, R. and Kopriva, S. (2002) Functional knockout of the adenosine 5'-phosphosulfate reductase gene in *Physcomitrella patens* revives an old route of sulfate assimilation. *Journal of Biological Chemistry,* 277, 32195-32201.

[52] Girke, T.; Schmidt, H.; Zahringer, U.; Reski, R. and Heinz, E. (1998) Identification of a novel delta 6-acyl-group desaturase by targeted gene disruption in *Physcomitrella patens. Plant Journal,* 15, 39-48.

[53] Bhatla, S. C.; Kiessling, J. and Reski, R. (2002) Observation of polarity induction by cytochemical localization of phenylalkylamine-binding receptors in regenerating protoplasts of the moss *Physcomitrella patens. Protoplasma,* 219, 99-105.

[54] Chinnasamy, S.; Bhatnagar, A.; Claxton, R.; Das, K. C. (2010) Biomass and bioenergy production potential of microalgae consortium in open and closed bioreactors using untreated carped industry effluent as growth medium. *Bio resources Technology,* 101, 6751-6760.

[55] Larrick, J. W. and Thomas, D. W. (2001) Producing proteins in transgenic plants and animals. *Current Opinion in Biotechnology,* 12, 411-418.

[56] Biefreund, N. M.; Reski, R. and Decker, E. L. (2003) Use of an inducible reporter gene system for the analysis of auxin distribution in the moss *Physcomitrella patens. Plant Cell Reports,* 21, 1143-1152.

[57] *Greenovation Pharmaceutical Protein Production in Moss.* 2013, October. Available from: www.greenovation.com/products.html

[58] *Greenovation Pharmaceutical Protein Production in Moss.* 2013, October. Available from: www.greenovation.com/services.html

In: Moss
Editor: Jorma Mohamed
ISBN: 978-1-63117-396-7

Chapter 4

ENDOPHYTIC BACTERIA ASSOCIATED WITH *SPHAGNUM* MOSSES: ECOLOGICAL DIVERSITY AND APPLICATION FOR AGRICULTURAL MICROBIOLOGY

A. V. Shcherbakov[1*], E. Yu. Kuzmina[2], E. D. Lapshina[3], A. Bragina[4], C. Berg[5], G. Berg[4], E. N. Shcherbakova[6], V. K. Chebotar[1] and I. A. Tikhonovich[1]

[1]All-Russia Research Institute for Agricultural Microbiology of the Russian Academy of Agricultural Sciences, St. Petersburg, Russia
[2]Komarov Botanical Institute of the Russian Academy of Sciences, St. Petersburg, Russia
[3]UNESCO department "Environmental dynamics and global climate changes" Yugra State University, Khanty-Mansiysk, Russia
[4]Institute of Environmental Biotechnology, Graz University of Technology, Graz, Austria
[5]Institute of Botany, University of Graz, Graz, Austria
[6]National University of Life and Environmental Science of Ukraine, Kiev, Ukraine

[*] E-mail address: avsherbakov@bisolbi.ru.

Abstract

Sphagnum mosses serve as a unique habitat for microorganisms, which play an important role both for the host plants and the peatland ecosystems. The aim of the present study was to isolate endophytic bacteria from the tissues of *Sphagnum* mosses and to screen them for strains promising for further application in agricultural microbiology. About 50 samples of *Sphagnum fallax* (H. Klinggr.) H. Klinggr. and *Sphagnum magellanicum* Brid. were collected in the Austrian Alps, Lenindgrad Region and West Siberia in 2009–2012. Endophytic bacteria were detected inside the moss plants using fluorescent in situ hybridization (FISH) followed by confocal laser scanning microscopy (CLSM). Altogether, more than 400 isolates were obtained by cultivation on the nutrient media. Examination of the isolates for the antagonistic activity revealed that more than 50% of them could suppress the growth of phytopathogenic and toxigenic fungi. More than 30% of isolates showed some antagonistic activity against microbial phytopathogens. The isolated strains could colonize crops and promote their growth. Molecular-genetic identification coupled with physiological/biochemical characterization showed that the dominant endophytic groups belonged to the genera *Burkholderia*, *Pseudomonas*, *Flavobacterium*, *Serratia* and *Collimonas*. The isolated endophytes were shown to be promising objects for the development of effective growth promoting and protective microbiological preparations to be used in agriculture.

Introduction

Sphagnum mosses are the major, often dominant component of wetland vegetation. These mosses have an almost cosmopolitan distribution, occurring in the tropics, where they grow in the mountains, in the temperate zones of both hemispheres and in the Arctic and the Subantartic. The subclass Sphagnidae is represented by the order *Sphagnales*, containing a single family *Sphagnaceae* with the only genus, *Sphagnum*. It comprises over 300 species, which are rather difficult to distinguish morphologically. The Russian flora contains 42 *Sphagnum* species. *Sphagnum* mosses are especially common in the temperate zone of the Northern hemisphere, where they form vast expanses of raised bogs. Owing to their highly specialized morphology and anatomy, *Sphagnum* species possess an outstanding capacity to absorb and retain moisture. Once settled in a wet habitat, they promote fast bogging of the area.

Endophytic colonization of *Sphagnum* hyaline cells by microorganisms was first observed by Stewart [1] and somewhat later described with the use of electron microscopy [2]. These authors observed vegetative cells and heterocysts of cyanobacteria inside *S. riparium* hyalocytes. The cyanobacteria formed aggregations and were embedded into slimy matrix. *Sphagnum* hyalocytes also contains heterotrophic bacteria, *Methanosarcina*-like bacteria and fungal mycelium. Based on the observation that *Sphagnum* mosses with cyanobacterial associates grow faster than those without them, Granhall and Hofsten [2] suggested a model of transfer of chemical elements between the host plan and the endophytic microbial associates in hyalocytes. According to this model, cyanobacteria interact not only with the host plant but also with other heterotrophic bacteria present in the confined space of the cell. Such heterotrophic bacteria may utilize nitrogen and carbon compounds produced by cyanobacteria, while bacterial respiration and nitrogen uptake stimulates nitrogen fixation, increase in CO_2 level and decrease in O_2 concentration. The products of nitrogen fixation such as hydrogen and ammonium may be used by methanogenic bacteria.

SPHAGNUM-ASSOCIATED METHANOTROPHY

The possibility that internal tissues of *Sphagnum* plants may be colonized by methanotrophic microorganisms was first discussed in detail by Raghoebarsing and colleagues [3,4]. These studies dealt with the symbiosis between *Sphagnum* mosses and methanotrophic bacteria ensuring carbon production for the plant construction. Dedysh et al. [5 - 7] suggested that methanotrophic bacteria were not simply present in the bog water or peat deposits but inhabited *Sphagnum* mosses, including their internal parts. Studies of the stem cortex of *Sphagnum cuspidatum* by fluorescent *in situ* hybridization and epiluminescent microscopy demonstrated that the inner part of hyaline cells was indeed inhabited by methanotrophic bacteria. Hybridization was performed in this case with the use of group-specific oligonucleotide probe Alf968, suitable for revealing the whole cluster of alphaproteobacteria. Characteristically, the bacteria did not simply float in the liquid in hyalocytes but were attached to their inner surface, forming microcolonies there.

As shown by 16S rRNA gene sequencing, the dominant methanotrophs among the bacterial washed off the surface of *S. cuspidatum* plants were *Methylocella palustris* and *Methylocapsa acidiphila*. These bacteria inhabited the internal tissues of *Sphagnum* plants. It was suggested that they were involved in symbiosis with *Sphagnum* plants, ensuring carbon assimilation from methane and its accumulation in wetland ecosystems in the form of plant tissues [6, 7].

According to different estimates, methane, oxidized by bacteria, is the source of up to 10–15% [3] or 10–30% [8] of carbon in the biomass of *Sphagnum* mosses. Kip et al. [9, 10], using pmoA gene sequencing and pyrosequencing, studied methanotrophic bacteria associated with internal tissues of *Sphagnum* plants. *Sphagnum* samples from different parts of wetlands in the Netherlands and Belgium were examined, including *S. cuspidatum* and *S. denticulatum*. The dominant components of endophytic microbial communities were methanogenic alphaproteobacteria from the genera *Methylocystis* and *Methylosinus*. It was also shown, by tracer analysis, that carbon from methane molecules was involved into plant metabolism, being accumulated in lipid inclusions [10]. This is a definite proof of symbiotic relationships between *Sphagnum* mosses and methanotrophs.

In the above studies [9,10] methanotrophs associated with *Sphagnum* mosses were referred to as "endophytes", despite the specific location of bacteria in *Sphagnum* tissues (in this case we are talking about the hyaline cells and intercellular tissue of *Sphagnum*). The presence of representatives of *Burkholderiales* and *Pseudomonaceae* spp. was also mentioned [2] but, as these microorganisms do not oxidize methane, they were outside the scope of the research. Gammaproteobacteria were mostly represented by *Methylomonas* and *Methylobacter*.

Field experiments in the tundra (Eastern Siberia) showed that methane emission into the atmosphere depended on the prevailing bog vegetation [11]. The emission was twice as low in *Sphagnum* bogs than in bogs where sedges and bushes were dominant. In the author's opinion, this phenomenon is associated with methane assimilation by endophytic bacteria in *Sphagnum* mosses.

A recently coined term "Sphagnum-associated methanotrophy" [8, 12] is applied to the stable symbiosis between *Sphagnum* mosses and methanotrophic bacteria colonizing their inner parts. The latter assimilate methane and provide mosses with carbon.

The studies by Larmola [8] and Putkinen [12] focussed on reactivation of methane oxidation after drought. The authors aimed to reveal how methane assimilation resumed during rising water table in the peatland ecosystem. Two hypothesis were proposed: that methanotrophs survived the drought inside the moss plants or that methanotrophs in the ground water actively colonized the moss plants as the water table level rose. Experiments were conducted on *S. rubellum* in a boreal peatland in the south of Finland. Three days after *Sphagnum* plants lacking methane-oxidizing bacteria were "transplanted" next to the plants that did possess them the "inactive" *Sphagnum* plants were colonized. Microarray analysis showed that the composition of methanotrophs in both groups was the same, with *Methylocystis*-like microorganisms being the dominant component.

In the second part of the study *S. rubellum* samples lacking methanotrophs were submerged into unfiltered flark water (with methanotrophs) or water sterilized by filtration. Methanotrophs present in the water actively colonized Sphagnum plants, increasing the number of pmoA gene copies more than 60 times.

The study of physiological and biochemical properties of the isolated strains of methanotrophic bacteria showed that almost all the strains could grow on nitrogen-free media and some strains could produce ethylene on media when oxygen content was 5-10%, indicating their potential for molecular nitrogen fixation [11].

CYANOBACTERIA ASSOCIATED WITH *SPHAGNUM* MOSSES AND THEIR ROLE IN N_2 ASSIMILATION

Associations between bryophytes and nitrogen-fixing microorganisms have long been known [13-15]. A special role in such associations is played by cyanobacteria [16]. Bryophytes, unlike lichens, lack specialized morphological structures for such associations, and cyanobacteria colonizing them either epiphytically or intercellularly, by penetration into dead hyaline cells [17].

Our knowledge of cyanobacteria associated with bryophytes mostly concerns polar and subpolar regions, where such associations are the major actors in the processes of molecular nitrogen fixation and ensure much of the ecosystems' productivity [17].

Cyanobacteria associated with bryophytes in the Antarctic has been studied in some detail [18 - 21]. Such associations, however, are rather poor in cyanobacterial taxa. The dominant species were *Phormidium frigidum* Fritsch and *Nostoc commune* Vaucher, with *Microcystis parasitica* Kutzing also being common [21]. *Nostoc* species were found as epiphytes on the leaves of mosses in the Arctic, alongside with *Oscillatoria* and *Anabena* [22]. Cyanobacteria appear to be transferred from plant to plant mostly passively, with the water currents [17] however, chemotaxis and active colonization have also been suggested [23].

We now know a lot about epi- and endophytic cyanobacterial colonization of moss plants. There are data on uneven colonization of mosses by epiphytic bacteria [24] and on nitrogen-fixing activity of associations between cyanobacteria and bryophytes [25].

A model of active intercellular colonization of *Sphagnum* plants by cyanobacteria has been proposed [14]. It is often noted that cyanobacteria are more abundant in the middle (green) part of the moss plant then in the upper part. Broady et al. [24] studied vertical micro-distribution of phototrophs (including three species of cyanobacteria) in moss plants for almost three years. These authors proposed a model of growth and life cycles of *Nostoc muscorum* in the moss shoot.

Early in summer thalli with heterocysts were found in the primary apical structures of the plant; then, as it grew, cyanobacterial trichomes fragmented and mobile hormogonia were freed and rose upwards with the growing moss shoot. In this way, the hormogonia reached the site of the ageing terminal bud, where heterocysts were formed. Young shoots contained no cyanobacterial associates, while in the ageing plant parts both the plant tissue and the cyanobacterial associates decayed, enriching the environment with the organic substances. It was suggested that *Nostoc muscorum* formed such specific mobile structures for supporting the population in the regions with a high solar activity level [24].

Water-filled hyaline cells of *Sphagnum* plants are microniches for cyanobacteria, colonization proceeding via the cell pores. *Nostoc* sp. was found in the tissues of *Sphagnum lindebergii* and *S. riparium* but not in *S. balticum*, *S. fuscum* and *S. annulatum* [2]. These authors suggested that the colonization mode, intercellular or epiphytic, was determined by ecosystem's pH. At low pH values (4.2-4.9), as in the habitats of *S. lindebergii* and *S. riparium*, cyanobacteria, "avoiding" the acidic environment, colonized the cells of Sphagnum plants. When pH was lower still (less than 3.8), as in the habitats of *S. balticum*, *S. fuscum* and *S. annulatum*, the activity of cyanobacteria was totally suppressed. At higher pH values, when the acidity was close to neutral, cyanobacteria colonized the mosses epiphytically or were free-living [26].

Solheim and Zielke [17] studied the distribution of cyanobacteria in bryophytes and their acetylene-reductase activity. Their results are similar to those of Broady et al. [24], Smith et al., [25] and others. Using epiluminescent and laser confocal microscopy, Solheim and Zielke [17] studied associations between cyanobacteria and different moss species in the Arctic (Svalbard, Norway) and the Subarctic (Abisko, Sweden). Cyanobacteria were mostly located along the leaves and sometimes, as in *S. uncinata*, also inside the cells. The best images of localized cyanobacteria were obtained with the use of confocal laser microscopy but the main difficulties of studying the localization of cyanobacterial associates were associated with the preparation and correct fixation of hollow moss stems. On the high resolution images cyanobacteria can be seen to occupy the space inside the matrix between the leaves and the stems. The matrix promotes the stability of the association and facilitates transfer of substances between the microorganism and the "macroorganism".

Fast assimilation of nitrogen by cyanobacteria and its transfer into the moss plant tissues was shown with the use of radioactively labelled molecular nitrogen [13]. Though the amounts of fixed molecular nitrogen in the Polar regions are much lower than in the temperate and tropical ones, it is nevertheless a considerable source of nitrogen in land ecosystems [13, 27]. Few nitrogen compounds enter such ecosystems with precipitation and its fixation by soil bacteria is also poor due to low temperatures and limited sources of nutrition in the soil [25, 28]. According to Dodds et al. [29], the contribution of cyanobacteria associated with bryophytes into assimilation of atmospheric nitrogen makes up 2-58% in the Arctic and 42-84% in the Antarctic. The removal of nitrogen from such ecosystems is compensated for by the activation of nitrogen fixation by cyanobacteria [30].

In temperate and tropical regions the contribution of cyanobacteria associated with bryophytes may be high in certain specific habitats. For instance, *Sphagnum* mosses were shown to harbour nitrogen-fixing cyanobacteria only in the open bog ecosystems [31]. In temperate coniferous forests they have no nitrogen-fixing cyanobacterial associates. Their nitrogenase activity is very low and ensured by heterotrophic bacteria. The rate of nitrogen assimilation by *Sphagnum*-associated cyanobacteria in the open bogs makes up 37.9 kg N ha^{-1} per year [14], depending heavily on humidity and light intensity [32].

The rate of atmospheric nitrogen fixation by cyanobacterial associates of mosses was often shown to be influenced by various abiotic factors [22, 33 - 43]. The major factors mentioned in these works were temperature, humidity, light intensity and nutrient availability.

Heterotrophic Bacteria Associated with *Sphagnum* Mosses

Colonization of *Sphagnum* hyaline cells by heterotrophic bacteria was first mention in the works of Swedish researchers [2]. Their electron microscopic studies revealed, alongside with cyanobacteria, also heterotrophic ones. The most complete body of information on the heterotrophic bacteria associated with *Sphagnum* mosses, including the data obtained by molecular-genetic methods, can be found in Opelt and Berg [44], Vandamme et al. [45], Opelt et al. [46] and Bragina et al. [47].

The studies quoted above originally concerned not only the associates of *Sphagnum* mosses but also those of other bryophytes, such as *Tortula ruralis* and *Aulacomnium palustre*, which are usual components of plant communities at the southern Baltic Sea coast [44]. Electron microscopy revealed single heterotrophic bacteria on the plants' surface. Molecular-genetic analysis showed that the dominant bacterial groups associated with *Sphagnum rubellum* were the genera *Pseudomonas, Serratia* and *Burkholderia*, while subdominant clusters were represented by *Acetobacter, Acidocella* and others. Inoculation on solid culture media revealed a vast diversity of *Sphagnum*-associated bacterial groups. The total abundance of heterotrophic bacteria isolated from *Sphagnum* tissues made up 10^5-10^6 CFU/gr of plant tissue.

Pseudomonas (P. putida, Pseudomonas sp.), Serratia (S. proteamaculans, S. liquefaciens), Burkholderia (B. phenazinium, B. fungorum) dominated among the bacteria isolated into pure cultures; species of *Collimonas, Xantomonas* and *Bacillus* were also present. Out of 83 strains studied for signs of antagonistic activity, 99% showed fungicidal activity against *Verticillium dahliae*.

With the use of SSCP method, the composition of microbial community was shown to vary depending on the moss species (two Sphagnum species, *Sphagnum magellanicum* and *S. fallax*, were studied) [46]. Samples of the same moss species from Germany and Norway had a similar composition of microbiota. The total of 1222 strains of heterotrophic bacteria were isolated from surface-sterilized *S. fallax* and *S. magellanicum* samples by inoculation on solid cultural media [48]. Molecular-genetic identification revealed species of *Burkholderia, Serratia, Bacillus, Paenibacillus, Pseudomonas, Rahnella* and *Dyella* (dominant genera) as well as some representatives of *Moraxella, Microbacterium* and *Streptomyces*. More than 26% of the isolated strains had fungicidal activity against phytopathogenic fungi *V. dahliae* and *Rhizoctonia solani. S. fallax* had a higher percentage of strains with fungicidal activity than *S. magellanicum*. A special place among the isolated strains of *Sphagnum*-associated microorganisms was occupied by various *Burkholderia* species, bearing nifH genes and apparently possessing nitrogen-fixing activity.

Fourteen *Burkholderia* strains from gametophytes of *Sphagnum* mosses were studied by multifactor taxonomy, including DNA-DNA hybridization and physiological-biochemical characteristic [45]. Based on these data, two new species were described, *Burkholderia bryophila* and *B. megapolitana*. According to the authors, they are a specific component of microbial communities associated with *Sphagnum* mosses.

Despite the high proportion of strains with antagonistic activity against fungal pathogens, the authors also report colonization of *Sphagnum* mosses by bacteria from the genera where most representatives are opportunistic pathogens of humans and other warm-blooded animals [49]. For example, among the bacterial strains isolated from the tissues of *S. fallax* and *S. magellanicum* there were representatives of the genera *Staphylococcus, Hafnia, Yersinia* and *Pantoea*. However, the authors give no information about whether the isolated strains were indeed human or animal pathogens.

Detailed molecular-genetic studies of the total DNA extracted from the same moss species (*S. fallax* or *S. magellanicum*) and the analysis of clone libraries showed that non-culturable species of *Alphaproteobacteria* and *Gammaproteobacteria* were dominant in the microbial communities of *S. magellanicum* [47]. Microbial community of *S. fallax* was very different, the dominant species belonging to *Verrucomicrobia* and *Planctomyces*. The authors analyse the dependence of the composition of specific microbial communities of *Sphagnum* mosses on abiotic factors such as nutrient availability and water pH. They suggest that microbial community composition may change with the changing environmental conditions.

Yeasts and Arbuscular Mycorrhizal Fungi Associated with Mosses

Sphagnum mosses are specific habitats for epiphytic yeasts [50]. The composition of the epiphytic yeast community of *Sphagnum* mosses is different from that of other vascular plants in the same ecotope. The average abundance of epiphytic yeasts was 1.5×10^3 CFU/gr of plant tissues in *Sphagnum girgensohnii* Russ., *S. magellanicum* Brid., *S. balticum* (Russ.)

C.Jens., and *S. angustifolium* (Russ.) [50]. Seasonal dynamics of abundance was observed, with minimum values in late autumn and winter and maximum ones, in spring [51].

Species composition of epiphytic yeasts of bog plants comprised more than 70 species [51]. The dominant species of *Sphagnum* mosses and vascular plants were the same. Two species, *Rhodotorula mucilaginosa* and *Cryptococcus magnus*, made up, on the average, about 50% of the abundance. However, the composition of minor species varied considerably. The main difference was that the mosses contained a vast array of various psychrophilic yeast species.

Symbiosis between mycorrhizal fungi and bryophytes were first described in the early 20th century [52] and later mentioned in several other studies [53-55]. Bryophytes were also reported to form arbuscular mycorrhizal symbiosis with lichens [56-58]. Hyphae and vesicles of nonseptate endophytic fungi, similar to the arbuscular mycorrhiza (AM) of higher plants, were first found in *Sphagnum* species by Butler [59].

Downing [60] described AM of *Endogone fasciculata* (=*G. fasciculatum*) fungi in bryophytes from Alberta wetlands (Canada). They were located in the ground as well as around the plant roots and inside (or around) the moss leaves. Rabatin [61] reported that *G. tenuis* formed AM with *Pogonatum* moss plants but did not observed any hyphae inside the plant tissues. Spores, hyphae and vesicles of *G. epigaeum* were found in the moss *Funaria hygrometrica* only when it grew near to its “companion” plant, the asparagus [62].

The structure of mycorrhizal symbiosis in different bryophytes has recently been studied in detail [63]. Out of 24 moss species from 16 families, spores, vesicles and intercellular and extracellular hyphae were found in 21 (88%) moss species. Contrary to the earlier descriptions [60-62], the AM fungi formed these structures inside the moss stems and leaves but not inside the rhizoids. This location is not surprising taking into account that the green leaves of mosses consist of non-specialized parenchymal tissue similar to the non-differentiated cortical tissue of roots of vascular plants. The fungi isolated from the moss plants belonged to the genus *Glomus* (11 species), *Acaulospora* (2 species), *Gigaspora* and *Paraglomus* (1 species each). *G. sinuosum* was found in 9 moss species (69%), *G. ambisporumin*, in 3 species, while each of the fungal species *G. fasciculatum, G. microaggregatum*, and *Glomus sp.* was found in two moss species.

PLANT GROWTH PROMOTING BACTERIA OF *SPHAGNUM* AND PROSPECTS OF THEIR USE IN AGRICULTURAL MICROBIOLOGY

In our work [47, 64-66] we studied two *Sphagnum* species occupying different econiches in bog ecosystems but often co-occurring in the same biotopes, *Sphagnum fallax* (H. Klinggr.) H. Klinggr. and *S. magellanicum* Brid. (species names are given after [67]). Both species have a wide distribution. A broad geographical scope of sampling provided information at the Eurasian scale (Table 1). The westernmost sampling region were the Austrian Alps. The samples were taken in the foothills, at three sites at a distance of 200 km from each other. The central sampling region was the Leningrad Region (Russia). This area is rich in wetlands and bogged areas. Samples from this region examined in most detail, including the study of ecological diversity of the *Sphagnum*-associated bacteria. The easternmost sampling region was the Khanty-Mansi Autonomous Area in Western Siberia. This regions is also rich in bogs, and both *S. fallax* and *S. magellanicum* occur there.

State-of-the-art molecular biological and confocal scanning laser microscopic methods made it possible to localize endophytic bacteria in the tissues of vegetating *Sphagnum* plants. FISH and CSLM allowed us to reconstruct a three-dimensional model of the structure of internal tissues inhabited by endophytic bacteria. Owing to a unique structure of vegetating parts of *Sphagnum* plants, which contain both living chlorophyll-bearing cells and dead hyaline cells, the tissues of these plants provide unique micro-niches for microbial communities, which form close symbiotic relationships with the host plants. This symbiosis appears to date back to the earliest stages of the land plants' evolution. *Sphagnum* mosses, or rather their evolutionary ancestors, were the first plants to colonize land. Therefore, modern *Sphagnum* mosses are a unique model for the study of such symbioses. These studies also affect our ideas about the functional properties of the *Sphagnum* tissues. Hyalocytes, with their spongy structure and large pores, were previously thought merely to store water and supply it to the plant as need be. Studies of the last decade, including those by our group, showed that the function of hyaline cells is not confined to water storage. They also provide "hosting" for the microbial communities, ensuring the functioning of a unique symbiosis.

Table 1. Geographical points of sampling *Sphagnum* mosses

Geographical region	Sampling point	Geographical coordinates	Date	Abiotic parametrs	Moss species	Strain abbreviator
Alps Region	"Rootmoos" (Austria, Styria)	N 47°41′ E 15°09′	02.09.2009	pH 4.3–4.5; 8°C	*S. magellanicum* *S. fallax*	AM1 AF1
	"Waasenmoos" (Austria, Salzburg)	N 47°18′ E 12°24′	03.09.2009	pH 3.3–3.7; 10°C	*S. magellanicum* *S. fallax*	AM2 AF2
	"Pürgschachen" Moor (Austria, Styria)	N 47°34′ E 14°20′	04.09.2009	pH 3.9–4.2; 8°C	*S. magellanicum* *S. fallax*	AM3 AF3
Saint-Peterburg Region	Bog "Polesye" (Russia, Vyborg Province)	N 60°44′ E 29°04′	19.07.2010	pH 4.2–4.6; 15°C	*S. magellanicum* *S. fallax*	RVM1 RVF1
	Voloyarvi Lake (Russia, Vsevologsk Province)	N 60°18′ E 30°48′	03.10.2009	pH 4.7–5.0; 5°C	*S. magellanicum* *S. fallax*	RM1 RF1
	Bog "Kardon Kirpichny" (Russia, Gatchina Province)	N 59°16′ E 29°54′	06.10.2011	pH 4.4–5.1; 7°C	*S. magellanicum* *S. fallax*	RSM1 RSF1
West Siberia	Bog "Chistoe" (Russia, Khanty-Mansiysk Region)	N 61°03′ E 69°27′	30.08.2012	pH 3.2–4.0; 18°C	*S. magellanicum* *S. fallax*	HM1 HF1
	Bog "Muhrino" (Russia, Khanty-Mansiysk Region)	N 60°53′ E 68°41′	31.08.2012	pH 4.1–5.0; 21°C	*S. magellanicum* *S. fallax*	HM2 HF2
	Bog "Kukushkino" (Russia, Khanty-Mansiysk Region)	N 60°57′ E 69°41′	01.09.2012	pH 4.4–5.0; 17°C	*S. magellanicum* *S. fallax*	HM3 HF3

A combination of FISH and CSLM and three-dimensional computerized imaging revealed a vast array of microorganisms inside *Sphagnum* tissues. The use of universal and group specific oligonucleotide probes allowed us to visualize representatives of Alpha-, Beta-, and Gammaproteobacteria (Figure 1). Bacterial microcolonies were also present on the outer surface of *Sphagnum* leaves and stems but it was inside the hyaline cells that their density was the highest. Three-dimensional computerized imaging showed that bacterial microcolonies and separate cells are tightly attached to the inner surface of hyalocytes, indicating their intimate interaction with the host plant.

Microbial communities' composition was studied in more detail in collaboration with our colleagues from Austria [47, 68]. Several common features of *Sphagnum* colonization by bacterial endophytes were revealed. The use of group-specific probes showed that Alphaproteobacteria were the dominant component of microbial communities in both *S. fallax* and *S. magellanicum*. In both these moss species Alphaproteobacteria were represented by various morphological forms: coccoid, rod-shaped, tetrads and *Sarcina*-like conglomerates. The second dominant component of *Sphagnum* microbial communities were Planctomyces. Their colonies were more often found in *S. fallax* than in *S. magellanicum*. Planctomyces were previously reported from peat communities of Sphagnum bogs as a subdominant component [69]. In our joint studies, however, we showed their ability to colonize the inner parts of *Sphagnum* plants.

Our research focussed on the bacterial genera the species of which are successfully used in agricultural microbiology. Therefore, oligonucleotide probes specific for Gamma- and Beta- proteobacteria were used for FISH. These bacterial classes are rich in genera and some of them are promising in respect to plant-microbial interactions. For instance, the class Betaproteobacteria includes the family Burkholderiaceae (order Burkholderiales), species of which were earlier found in associations with *Sphagnum* mosses [45]. Another promising genus revealed in our studies is *Collimonas* from the family Oxalobacteraceae (see below). Representatives of *Pandoraea*, *Delftia*, *Acidovorax*, *Janthinobacterium* (the order Burkholderiales) were isolated from bacterial associations in *Sphagnum* mosses. The class Gammaproteobacteria also contains bacteria capable of effective interactions with plants and enhancing their growth and development. It includes *Pseudomonas*, the largest PGPR bacterial genus. Some *Pseudomonas* species, the most effective colonists of higher plants, are used in biopreparations. The order Enterobacteriales mostly contains pathogenic and opportunistic species, but some its representatives (e.g. species of *Serratia*,

Klebsiella, *Rahnella*) may play the role of growth-stimulating and biocontrol agents.

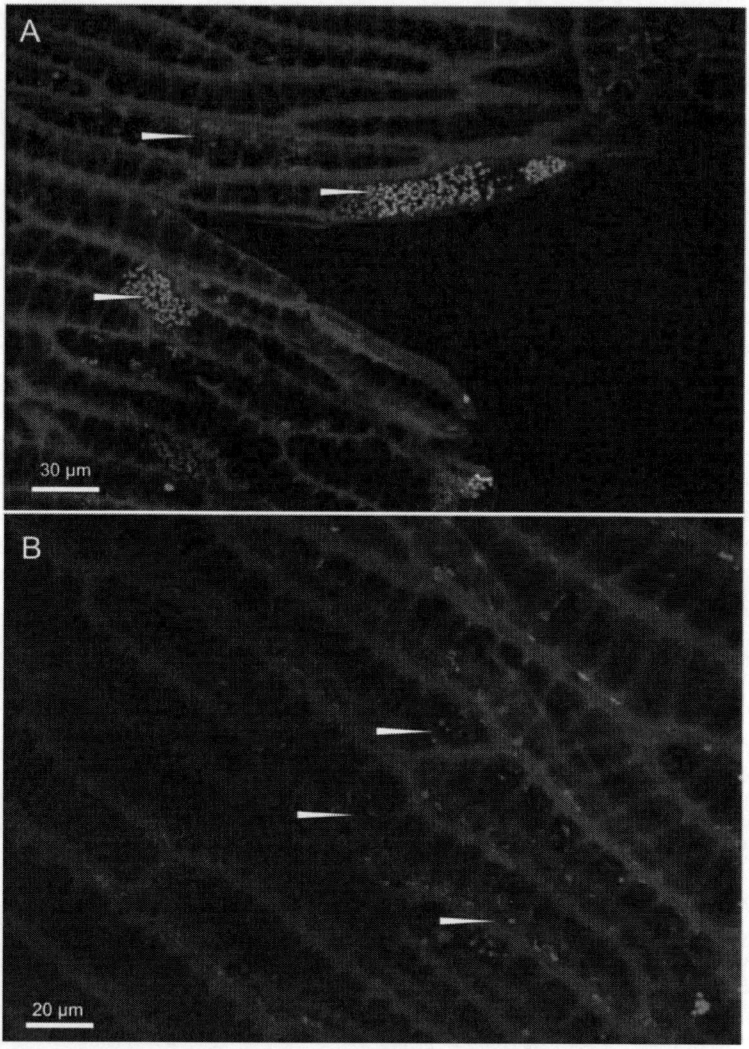

Figure 1. Localization of bacteria in *Sphagnum* gametophytes. Fluorescent *in situ* hybridization of branch leaves with universal bacterial probes EUB338, EUBII338, EUBIII338 (white arrows point to representatives of Eubacteria). Confocal laser scanning microscopy images. Scale bar: 30 µm and 20 µm.

Altogether, more than 400 culturable bacterial strains were isolated from the tissues of *S. fallax* and *S. magellanicum* sampled in the Austrian Alps, the Leningrad Region and Western Siberia. Their cultural and morphological properties were characterized. Most isolated bacteria (> 98%) were Gram-negative. They were represented by very small (< 1.5 μm) rounded or oval cells or short rods. On the R2A medium they formed fast- or slowly growing colonies, flattened and creeping, transparent or semi-transparent, brightly coloured (red, violet, pink, yellow, orange) or colourless (beige or milky white). The total abundance of microorganisms isolated onto culture media varied in the range 10^5-10^6 CFU/gr of plant tissue. Nitrogen-fixing and oligonitrophilic bacteria grew rather abundantly on the nitrogen-free medium but formed few morphotypes (not more than 5). They formed transparent colourless slimy colonies of middle size, either creeping on the agar surface or rounded and convex.

A characteristic feature of the studied isolates was an almost total absence of Gram-positive spore-forming *Bacillus* bacteria. Only two *Bacillus* strains were isolated. This may be associated with the excessive moistening in *Sphagnum* habitats or with the evolutionary history. Spore-forming bacteria probably emerged already after the bases of symbiosis between *Sphagnum* mosses and Gram-negative microorganisms had been laid. Formation of spores in bacteria is an adaptation to unfavourable environmental conditions such as lack of moisture and the drying up of the cells. A plant evolving towards life of land and still mostly submerged is more likely to form symbiosis with the microorganisms of the water environment, such as certain Gram-negative bacteria.

Molecular-genetic identification based on 16S rRNA gene sequencing with the use of BLAST and RDP revealed a high diversity of the isolated microorganisms. Dominant bacterial groups among the studied *Sphagnum* associates belonged to the genera *Pseudomonas*, *Serratia*, *Burkholderia*, *Flavobacterium*, *Collimonas* [65].

The study of fungicidal activity in more than 400 bacterial isolates from *Sphagnum* mosses revealed a high proportion of strains capable of suppressing the growth of phytopathogenic fungi (about 60%). More than 30% of the isolated strains to some extent suppressed the development of a number of major bacterial phytopathogens. Several strains were characterized by a complex activity against both fungal and bacterial pathogens, while some others had only one pronounced effect, bactericidal or fungicidal (Table 2).

Table 2. Taxonomic position, beneficial properties and physiological characteristics of the most promising strains of endophytic bacteria associated with *Sphagnum* mosses

Strain	Fungicidal activity, total	Bactericidal activity, total	Fermentative activity	P-solubilising ($Ca_3(PO_4)_2$/ phosphorite powder)	Growth on N-free medium	Level of auxin production (on L-tryptophan containing medium), µg/ml	Temperature optimum	Growth at +37	Growth at +4		pH optimum
Pseudomonas brenneri RM14B	+++	++	P(+), L(+)	+/+	-	0,11	28	-	+		6.0
Pseudomonas poae RF11D	++	+++	P(++)	+/+	-	0,04	28	-	-		6.0
Pseudomonas asplenii RF13H	+++	++	A(+++), L(++)	-/-	-	4,11	28	-	+		6.0
Serratia plymuthica AF24A	++	++	P(++), A(+), L(++)	-/-	-	5,43	28	-	+		7.0
Pseudomonas poae RM13C	+++	+	P(+), A(+)	+/+	-	-	18	-	-		5.0
Pseudomonas. sp. RM13D	+++	+	P(++)	+/+	-	0,13	28	-	-		6.0
Pseudomonas fluorescens RF14J	++	+	P(++), L(+)	+/+	-	6,23	28	-	+		6.0
Flavobacterium sp. RM14A	-	+++	P(++), A(+++), L(++)	-/-	-	-	18	-	+		7.0
Flavobacterium sp. RF14A	-	+++	P(+), A(++), L(++), C(++)	-/-	-	-	28	-	-		5.0
Burkholderia bryophila AM32A	-	-	A(+)	-/-	+	35,10	28	-	-		4.5
Burkholderia phenazinum RM13B	++	-	A(+)	-/-	-	-	28	-	-		6.0
Collimonas sp. RM14C	+	++	P(+), L(+)	-/-	+	1,23	18	-	-		6.0
Stenotrophomonas rhizophila RF13C	-	-	P(++), A(++)	-/-	-	-	28	-	±		6.0
Pseudomonas sp. RF14D	++	-	P(+), A(+)	-/-	-	8,10	28	-	-		6.0
Pseudomonas poae RF14H	++	+	-	+/+	-	-	28	-	+		6.0
Chryseobacterium soli RM12F	+	-	P(++), A(+++), L(+)	-/-	-	5,82	28	-	-		6.0

A number of strains with well-expressed fungicidal and/or bactericidal properties was selected for further research. Potentially valuable strains were further screened to reveal their capacity to produce plant hormones and stimulate the growth of agricultural crops. We used a simple method, based on the reaction with Salkovsky reagent, allowing to test simultaneously many strains for their capacity to produce indole acetic acid (IAA). A number of active isolates was thus revealed (Table 2). Visual assessment of the results of this qualitative reaction showed that some strains produced IIA or its derivatives rather actively on the L-tryptophan-containing medium: the staining was more intense than in the case of the control strain *Pseudomonas chlororaphis* 228. Besides, 12 out of the 18 studied strains (66%) showed positive results in qualitative reaction for IIA production. These strains were selected for the analysis of the auxin production by high-performance liquid chromatography (HPLC).

All the selected strains could produce the three major auxins: indole-3-lactic acid (ILA), indolyl-3-carboxylic acid (ICA) and indolyl-3-acetic acid (IAA). The most effective producers of IAA were *P. aspleni* RF13H (4.1 µg/ml), *Cryseobacterium sp.* RM12F (5.8 µg/ml), *Serratia plymuthica* AM24A (5.4 µg/ml), *P. fluorescens* RF14J (6.2 µg/ml), P. sp. RF14D (8.1 µg/ml). An especially active auxin producer was, however, *Burkholderia bryophila* AM31A. Its total auxin production reached 41 µg/ml, the degree of L-tryptophan conversion reaching 69.3%. Comparison of our data with the data on IAA production by some other PGPR strains [70] shows that some of our strains had a higher level of production of this major plant hormone. For instance, *P. chlororaphis* SPB1217 and *P. fluorescens* SPB2137 accumulated 0.96-1.88 µg/ml of IAA [70], whereas our strains accumulated, under the same experimental conditions, three times as much.

Testing the potential ability of bacterial strains isolated from *Sphagnum* mosses to colonize the rhizosphere of agricultural crops was the key issue of our research. This ability is the basis for the effectiveness of potentially promising strains under conditions of the system "higher plant–bacterium". To test this ability, the method of gnotobiotic systems was used.

The results indicate the positive answer. Colonization potential of 5 bacterial strains isolated from *Sphagnum* mosses was rather high [65]. The strains under study took well both in the rhizosphere and on the roots of wheat as well as on the surface of tomato roots. Their parameters were as good as those of the control strain, *P. chlororaphis* RR228, sometimes even exceeding them. *Serratia plymuthica* AM24A and *Stenotrophomonas rhizophila* RF13C had the maximum values of establishment on roots, both of wheat and tomatoes. Relative abundance of these bacteria on the roots as compared to the rhizosphere was also maximum, reaching 14%. The abundance of other strains was also high, comparable with the colonization activity indices of the control strain. The average abundance of all the strains on tomato roots was several times greater than on wheat roots.

Epiluminescent miscroscopy showed that strains of *Serratia plymuthica* AM24A, *Flavobacterium saccharophilum* RM14A, *Pseudomonas brenneri* RM14B, *Pseudomonas poae* RF11D and *Collimonas* sp. RM14C actively colonized the surface of tomato roots [64, 65]. They formed microcolonies or biofilms directed along the root axis. A characteristic feature of the microcolonies was that they were located at the junctions of rhizodermal cells. The bacterial colonies were elongated along the longitudinal root axis and lay rather deep at the sites where the plant cells touched each other. Single bacterial cells were also found on the surface of root hairs. Their number there was much lower than directly on the rhizodermal surface.

Growth-stimulating effect of two strains, *Serratia plymuthica* AM24A and *Pseudomonas asplenii* RF13H, was demonstrated in a micro-vegetation experiment on radish plants *Raphanus sativus* L. var. radicula in 2011. Treatment of radish seeds with the bacterial suspension with a titre of 10^7 CFU/ml had a positive effect on the plant growth and development: the average weight of both the root crops and the herbage was increased. For example, *P. asplenii* RF13H ensured an increase in the average raw weight of the root crops by 25% and *S. plymuthica* AM24A, by 40% [64]. In 2012, growth-stimulating effect of a greater number of strains from diverse taxonomic groups was studied [65]. Growth-stimulating effect, expressed in the enhanced herbage development, was revealed in a micro-vegetation experiment on tomato plants *Solanum lycopersicum*. *Pseudomonas brenneri* RM14B, *P. asplenii* RF13H and *P. fluorescens* RF14J were the most effective in this experiment, ensuring a greater increase in the herbage weight than the control strain *P. chlororaphis* RR228.

In an experiment on the biocontrol of wheat root rot induced by *Fusarium culmorum*, inoculation of wheat seeds with *P. asplenii* RF13H decreased the susceptibility of plant to this disease from 90 to 40% [65]. This disease was expressed in the darkening of some root areas and a browning of the stem bases as well as the damage or death of the root system. As noted in the experimental variant without the pathogen, about 50% of the seeds had an internal infection (presumably by *Alternaria* sp.), which was not eliminated by surface sterilization of seeds. It was expressed in the blackening of the kernel in the control plants. Inoculation by the *P. asplenii* strain suppressed completely the development of the internal seed infection in the variant without the additional pathogen.

Our research showed for the first time that bacteria isolated from *Sphagnum* mosses can colonize agricultural crops and enhance their growth and development. We isolated several promising strains which can actively colonize the rhizosphere and the rhizoplane of agricultural crops. Their abundance was comparable with that of the control strain used for production of a microbiological preparation. The promising strains had a high antagonistic potential against the bacterial and fungal plant pathogens. The first time was selected strains with an expressed growth-stimulation effect on the agricultural crops and demonstrated their effectiveness in micro-vegetative experiments. Strains with phosphate-mobilizing properties also had a growth-stimulating effect under conditions of soil enrichment with additional phosphorus sources poorly accessible to plants. Our results indicate a high potential of microbial preparations based on endophytic bacteria associated with *Sphagnum* mosses for the agricultural industry.

CONCLUSION

Studies of microorganisms associated with *Sphagnum* mosses commenced as soon as microbiology emerged as a science. In the early 20th century the first endosymbiosis between mycorrhizal fungi and *Sphagnum* plants were described. Somewhat later bacteria inside *Sphagnum* tissues were discovered. The advent of electron microscopy in the 1970s made it possible to reveal blue-green algae and heterotrophic bacteria inside the hyaline cells of mosses. At that time the researchers referred to such inhabitants as to "intercellular" algae or bacteria, since the term "endophyte" had not yet been coined.

Peatlands are quite extreme, nutrient-poor biotopes, yet *Sphagnum* mosses flourish there. What is it that helps them survive and achieve a high biomass?

It was only recently that the crucial role of microorganisms associated with *Sphagnum* mosses in the functioning of both moss plants and bog ecosystems on the whole was discovered. A hypothesis postulating an intricate relationship between methanogens, methanotrophs and *Sphagnum* mosses was suggested. According to it, methane produced by methanogens is assimilated by methanotrophs under aerobic conditions inside the hyaline cells of mosses; carbon thus fixed is provided to the moss plant itself. A similar symbiosis appears to exist between *Sphagnum* plants and cyanobacteria: the latter, residing in intercellular spaces, assimilate molecular nitrogen and provide it to moss plants. In the most recent studies nitrogen-fixing Eubacteria associated with *Sphagnum* mosses were also described [68].

Thus, a sound theory of symbiosis between *Sphagnum* mosses and associated microorganisms is now available. The study object of the thesis presented here are aerobic heterotrophic PGP-bacteria associated with *Sphagnum* mosses. Based on the literature and the original research by our group, the author proposed the following model of structural-functional interactions between *Sphagnum* plants and endophytic microorganisms in their tissues (Figure 2).

According to the suggested scheme, aerobic and microaerophilic heterotrophic bacteria from the PGP group occupy a key position in the system of symbiotic relationships between endophytic bacteria and host plants. The main functions of this group of microorganisms appear to be the production of plant hormones and vitamins for the host plant and antagonistic activity against phytopathogens.

For the first time in scientific record we studied microorganisms associated with *Sphagnum* mosses from several geographically distant regions of Eurasia. *Sphagnum* mosses were shown to provide a specific habitat for the so-called endophytic bacteria, which presumably play an important role not only for host plants but for the functioning of bog ecosystems on the whole. A unique collection of endophytic bacteria associated with *Sphagnum* mosses was made. Their cultural-morphological, physiological-biochemical and agricultural properties were studied. More than 50% of heterotrophic bacteria from *Sphagnum* tissues had pronounced antagonistic properties against a number of phytopathogens, which may be the basis of the ability of *Sphagnum* mosses to withstand bacterial and fungal diseases. Molecular-genetic identification based on 16S rRNA gene sequencing combined with the physiological-biochemical characteristic showed that the dominant bacterial groups among the studied *Sphagnum*-associated microorganisms belonged to the genera *Pseudomonas*,

Serratia, *Burkholderia*, *Flavobacterium*, *Collimonas* and some others. The same *Sphagnum* species sampled in different geographical regions were shown to provide habitats to the same taxonomic groups. A number of promising strains with valuable agricultural properties was selected for further studies. We showed for the first time that some endophytic bacteria isolated from Sphagnum can actively colonize agricultural crops, enhancing their growth and development. An important basis for the practical use of *Sphagnum*-associated microorganisms was thus laid. In particular, they may prove useful for the development of highly effective agricultural biopreparations.

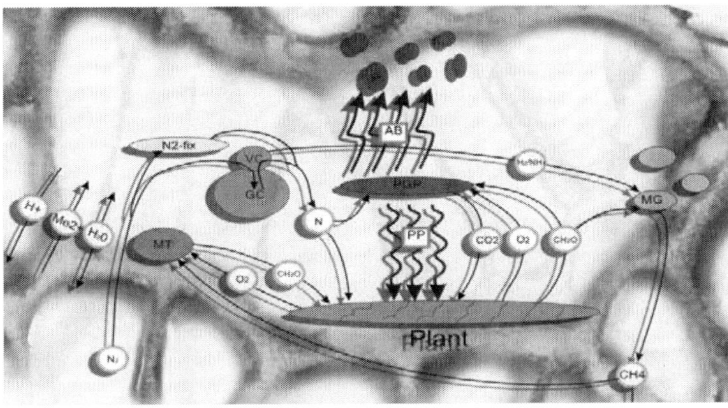

Figure 2. Structural and functional model describing the system of biochemical interactions between endophytic bacteria and plant moss: PGP-bacteria of group «Plant-Growth Promotion»; MG-methanogenic bacteria; MT-methanotrophic bacteria; VC-vegetative cells of cyanobacteria; GC-heterocysts of cyanobacteria; N_2-fix - nitrogen-fixing Eubacteria; P - phytopathogenic bacteria and fungi; Plant - host plant; AB - antibiosis; PP-phytohormons and vitamins production. The model is based on the system proposed Grabhell and Hofsten [2].

REFERENCES

[1] W. D. P. Stewart, *Nitrogen Fixation in Plants*, Athlone Press, London (1966).

[2] U. Granhall, A. Hofsten, *Physiol. Plant*. 36 (1976).

[3] A. A. Raghoebarsing, A.J.P. Smolders, M.C. Schmid, W.I.C. Rijpstra, M. Wolters–Arts, J. M. Derksen, *Nature*. 436 (2005).

[4] A. A. Raghoebarsing, A. Pol, K. T. van de Pas–Schoonen, A. J. Smolders, K. F. Ettwig, W. I. Rijpstra, S. Schouten, J. S. Damsté, H. J. Op den Camp, M. S. Jetten, M. Strous, Nature. 7086 (2006).

[5] S. N. Dedysh, N.S. Panikov, J.M. Tiedje, *Appl. Environ. Microbiol*. 64 (1998).

[6] S. N. Dedysh, W. Liesack, V.N. Khmelenina, N.E. Suzina, Y.A. Trotsenko, J.D. Semrau, A.M. Bares, N.S. Panikov, J.M. Tiedje, Int. J. Syst. Evol. Microbiol. 50 (2000).

[7] S. N. Dedysh, Microbiology. 78 (2009).

[8] T. Larmola, E. S.Tuittila, M.Tiirola, H. Nykanen, P. J. Martikainen, K. Yrjala, T. Tuomivirta, H. Fritze, *Ecology*. 90 (2010).

[9] N. Kip, B. E. Dutilh, Y. Pan, L. Bodrossy, K. Neveling, M. P. Kwint, M. S. M. Jetten, H. J. M Op den Camp, *Environ. Microbiol.* 3 (2011).
[10] N. Kip, J. F. van Winden, Y. Pan, L. Bodrossy, G. Reichart, A. J. P. Smolders, M. S. M Jetten, J. S. Damste, H. J. M. Op den Camp, *Nat. Geosci.* 3 (2010).
[11] F. Parmentier, J. van Huissteden, N. Kip, H. Op den Camp, M. Jetten, T. Maximov, *Biogeosciences,* 8 (2011).
[12] A. Putkinen, T. Larmola, T. Tuomivirta, H. Siljanen, L. Bodrossy, E. Tuittila, *Front. Microbiol.* 3 (2012).
[13] V. Alexander, M. Billington, D. Schell, *Subarct. Res. Sta.* 11 (1974).
[14] U. Granhall, T. Lindberg, *Ecol. Bull.* 26 (1978).
[15] K. Basilier, U. Granhall, T. Stenström, *Oikos.* 31 (1978).
[16] F. Gentili, M. C. Nilsson, O. Zackrisson, T. H. DeLuca, A. Sellstedt, *J. of Exp. Bot.* 56 (2005).
[17] B. Solheim, M. Zielke, *Cyanobacteria in Symbiosis,* Kluwer Academic Publishers, Dordrecht (2003).
[18] P. A. Broady, *Antarct. Rec.* 7 (1987).
[19] S. Ohtani, H. Kanda, *Polar Biol.* 1 (1987).
[20] B. Fumanti, P.Cavacini, S. Alfinito, *Polar Biol.* 7 (1997).
[21] S. Alfnito, B. Fumanti, P. Cavacini, *Nova Hedwigia.* 66 (1998).
[22] D. C. Jordan, P. J. McNicol, M. R. Marshall, *Can. J. Microbiol.* 24 (1978).
[23] A. N. Rai, E. Söderback, B. Bergman, *New Phytol.* 147 (2000).
[24] P. A. Broady, *Br. Antarct. Surv. Bull.* 48 (1979).
[25] V. R. Smith, *Appl. Environ. Microbiol.* 48 (1984).
[26] U. Granhall, H. Selander, *Oikos.* 24 (1973).
[27] D. M Chapin., C. S. Bledsoe, *Arctic Ecosystems in a Changing Climate,* Academic Press Inc., San Diego (1992).
[28] B. Solheim, A. Endal, H. Vigstad, *Polar Biol.* 16 (1996).
[29] W. K. Dodds, D. A.Gudder, D. Mollenhauer, *J. Phycol.* 31 (1995).
[30] D. R. Bazely, R. L. Jefferies, *J. Ecol.* 77 (1989).
[31] K. Basilier, *Lindbergia.* 5 (1979).
[32] K. Vlassak, E. A. Paul, R. E. Harris, *Plant Soil.* 38 (1973).
[33] A. Davey, *Polar Biol.* 2 (1983).
[34] A. Davey, H. J. Marchant, *Phycotogia,* 22 (1983).
[35] P. Christie, *J.Ecol.* 75 (1987).
[36] T. Nakatsubo, Y. Ino, *Ecol. Res.* 2 (1987).
[37] D. M. Chapin, L. C. Bliss, L. J. Bledsoe, *Can. J. Bot.* 69 (1991).
[38] R. Lennihan, D. M. Chapin, L.G. Dickson, *Can. J. Bot.*72 (1994).

[39] T. Liengen, R. A. Olsen, *Can. J. Microbiol*. 43 (1997).
[40] T. Liengen, *Can. J. Microbiol*. 45 (1999).
[41] L. G. Dickson, *Arc. Antarc. Alp. Res*. 32 (2000).
[42] B. Solheim, U. Johanson, T.V.Callaghan, J. A. Lee, D. Jones, L.O. Bjorn, *Oecologia*, 133 (2002).
[43] M. Zielke, A. S. Ekker, R. A. Olsen, S. Spjelkavik, B. Solheim, *Arct. Antarct. Alp. Res*. 34 (2002).
[44] K. Opelt, G. Berg, *Appl. and Envir. Microbil*. 70, 11 (2004).
[45] P. Vandamme, K. Opelt, N. Knochel, C. Berg, S. Schonmann, E. Brandt, L. Eberl, E. Falsen, G. Berg, *Int. J. of System. and Evol. Microbiol*. 57 (2007).
[46] K. Opelt, C. Berg, S. Schonmann, L. Eberl, G. Berg, *The ISME Journal*, 1. (2007).
[47] A. Bragina, C. Berg, M. Cardinale, A. Shcherbakov, V. Chebotar, G. Berg, *The ISME Journal*, 4, 6 (2012).
[48] K. Opelt, V. Chobot, F. Hadacek, S. Schönmann, L. Eberl, G. Berg, *Environ. Microbiol*. 9, 11 (2007).
[49] K. Opelt, C. Berg, G. Berg, *FEMS Microbiol Ecol*. 1, 61 (2007).
[50] A.V. Kachalkin, A. M. Glushakova, A. M. Yurkov, I. Yu. Chernov, *Microbiology*, 4, 77 (2008).
[51] A. V. Kachalkin, *Author's abstract* (2010).
[52] M. C. Rayner, *Mycorrhiza*, Wheldon and Wesley (1927).
[53] A. P. Kelley, *Mycotrophy in plants*, Chronica Botanica, Waltham, MA (1950).
[54] J. W. Gerdemann, *Phytopathol*. 6 (1968).
[55] J. L. Harley, *The biology of mycorrhiza*, Leonard Hill, London (1969).
[56] K. Turnau, M. Ronikier, *J. Acta. Soc. Bot. Pol*. 68 (1999).
[57] A. Schüßler, *Mycorhiza*, 10 (2000).
[58] J. Russell, S. Bulman, *New Phytol*. 165 (2005).
[59] E. J. Butler, *Trans. Br. Mycol. Soc*. 22 (1939).
[60] E. S. Dowding, *Trans. Br. Mycol. Soc.* 42 (1959).
[61] S. C. Rabatin, *Mycologia*, 72 (1980).
[62] J. L. Parke, R.G. Linderman, *Can. J. Bot*. 58 (1980).
[63] Y. Zhang, L. Guo, *Mycorrhiza*, 17 (2007).
[64] A. V. Shcherbakov, A. V. Krikovtseva, E. Yu. Kuzmina, C. Berg, N. V. Malfanova, M. Cardinale, G. Berg, V. K. Chebotar, I. A. Tikhonovich, *IOBC/WPRS Bulletin*, 78 (2012).

[65] A. V. Shcherbakov, A. V. Bragina, E. Yu. Kuzmina, C. Berg, A. N. Muntyan, N. M. Makarova, N. V. Malfanova, M. Cardinale, *Microbiology*, 3, 82 (2013).

[66] A. V. Shcherbakov, I. V. Rusakova, O. V. Orlova, N. I. Vorobyev, O. V. Sviridova, E. N. Shcherbakova, V. K. Chebotar, *Russ. J. Agricul. Biol.*, 1 (2014).

[67] M. S. Ignatov, O. M. Afonina, E. A Ignatova, *Arctoa,* 15 (2006).

[68] A. Bragina, S. Maier, C. Berg, H. Muller, V. Chobot, F. Hadacek, *Front. Microbiol.* 2, 275 (2012).

[69] S. N. Dedysh, T. A. Pankratov, S. E. Belova, I. S. Kulichevskaya, W. Liesack, *Appl. Environ. Microbiol.* 72 (2006).

[70] L. V. Kravchenko, N. M. Makarova, T. S. Azarova, N. A. Provorov, I. A. Tichonovich, *Microbiology*, 4, 71, (2002).

In: Moss
Editor: Jorma Mohamed
ISBN: 978-1-63117-396-7

Chapter 5

MOSS BIOMONITORING IN USE: SMALL SCALE AREA INVESTIGATION OF HEAVY METALS AIR POLLUTION - MINES AND SMELTER PLANT ENVIRONMENTS IN THE REPUBLIC OF MACEDONIA

Trajče Stafilov[1*], Biljana Balabanova[2], Robert Šajn[3] and Katerina Bačeva[1]

[1]Institute of Chemistry, Faculty of Science, Sts. Cyril and Methodius University, Skopje, Macedonia
[2]Faculty of Agriculture, Goce Delčev University, Štip, Macedonia
[3]Geological Survey of Slovenia, Ljubljana, Slovenia

ABSTRACT

Application of several moss species for monitoring of the anthropogenic impact on heavy metals air pollution in a small scale area was studied. Mosses were reviewed for their potential to reflect metal air pollution. The attention was focused on their quantification ability, underlying the metal accumulation in the moss plant tissue. Potential "hot spots" were selected in areas of lead and zinc mines (case study at

[*] Corresponding author: Email: trajcest@pmf.ukim.mk.

Zletovo mine and Sasa mine), copper mine (Bučim mine), ferronickel smelter plant (FENI industry) and the abandoned As-Sb-Tl mine at the Allchar area, as the main metal pollution sources in the Republic of Macedonia in the period between 2010-2012. There is continuous distribution of dust from ore, flotation tailings and slag surface. This results in air-introduction and deposition of higher contents of certain metals. Several moss species (*Hypnum cupressiforme*, *Campothecium lutescens, Scleropodium purum* and *Homolothecium sericium*) were used as plant sampling media. Determination of chemical elements was conducted by using both instrumental techniques: atomic emission spectrometry with inductively coupled plasma (ICP-AES) and mass spectrometry with inductively coupled plasma (ICP-MS). Combination of multivariate techniques (PCA, FA and CA) was applied for data processing and identification of elements associated with lithogenic or anthropogenic origin. Spatial distribution maps were created for determination and localization of narrower areas with higher contents of certain anthropogenic elements. In this way, influences of selected human activities in local (small scale) air pollution cases can be determined. Summarized data reveal real quantification of elements distribution not only in order to determine the hazardous elements distribution, but also present complete characterization of elements deposition in mines/smelter plant environs.

Keywords: Air pollution, moss, biomonitoring; ICP-AES, ICP-MS, heavy metals, spatial distribution, Fe-Ni smelter plant, Cu mine, Pb-Zn mine, As-Sb-Tl mine

INTRODUCTION

Environmental science is younger than the other disciplines in the natural sciences. In a relatively short period of time, certain new approaches and methods were introduced in association with the advances and new environmental applications. The environmental pollution at hazardous levels to life presents a global problem, and a macro-case for monitoring. Certain sub-disciplines have occurred over time in order to consider the realistic environmental conditions. Arguably, the understanding of atmospheric pollution is one of the most important emerging areas of environmental science. Atmospheric pollution represents solutions or suspensions of minute amounts of harmful compounds in the air (Valero, 2008). The degree and the extent of environmental changes over the last decades has given a new urgency and relevance for detection and understanding of environmental changes, due to human activities, which have altered global biogeochemical cycling of heavy metals and other pollutants (Hock and Seifert, 2003; Blagnytė and Paliulis, 2010; Glukhov and Prokhorova, 2011). Monitoring toxic air pollutants is needed for understanding their spatial and temporal distribution and ultimately to minimize their harmful effects. In addition to direct physical and chemical methods of air pollution monitoring, *bioindication* has also been used to evaluate air pollution risk (Valero 2008).

Heavy metals present only a part of the plurality of harmful compounds in the air. The degree of metals extent and distribution in the air depend on the emissions frequency (Longchurst and Brebbia 2013). However, higher contents of certain heavy metals introduced in the air create hazardous conditions both for the population and the environment. Air pollution with heavy metals is a global problem, but the hot spots occur and have influence on local level (Harmens et al., 2010). For this reason, a double type of monitoring program should be applied. The first one should cover larger areas, locating the hot spots in the investigated region; and then the smaller area where the local emissions of sources of heavy metals directly influence the local population and its environment.

1.1. The Term of Biomonitor/Bioindicator

Bioindicators include biological processes, species, or communities used to assess the quality of the environment and how it changes over time (Aboal et al., 2010). Changes in the environment are often attributed to anthropogenic disturbances (e.g., pollution, land use changes) or natural stressors (e.g., drought, late spring freeze), although anthropogenic stressors form the primary focus of bio-indicator research. The widespread development and application of bio-indicators has occurred primarily since the 1960s (Rühling and Tyler, 1968). Over the years, the repertoire of bio-indicators was expanded to assist us in studying all types of environments (i.e., aquatic and terrestrial) using all major taxonomic groups. However, not all biological processes, species, or communities can serve as successful bio-indicators. Physical, chemical, and biological factors (e.g., substrate, light, temperature, competition) vary among the environments (Fernández et al., 2007, 2012).

In common usage, the terms "biomonitoring" and "bioindication" are interchangeable. However, in the scientific community these terms have more specific meanings (Fränzle, 2006). Bioindicators qualitatively assess biotic responses to environmental stress, while biomonitors quantitatively determine a response. Hereinafter, the term "bioindicator" is used as a collective term which refers to all terms relating the detection of biotic responses of the environmental stress. Within this framework, there are three main functions of bioindicators: 1) to monitor the environment (physical and/or chemical changes), 2) to monitor ecological processes, and 3) to monitor biodiversity (Onianwa, 2001).

Examples of environmental, ecological, and biodiversity indicators can be found in many different organisms which inhabit many different environments. Lichens (a symbiosis among fungi, algae and/or cyanobacteria) and bryophytes (mosses and liverworts) are often used to assess air pollution. Bryophytes serve as effective bioindicators of air quality because they do not have roots, cuticle, and they acquire all their nutrients from direct exposure to the atmosphere (Wolterbeek, 2002). Their high surface area to volume ratio further encourages the interception and accumulation of contaminants from the air. The numerous benefits of bioindicators have spurred legislative mandates for their use in countries' monitoring programmes around the world. Yet, bioindicators are not without any problems. Like the canaries in the coal mine, we rely on the sensitivity of some bioindicators which function as early-warning signals. In some instances, we cannot discriminate natural variability from the changes of the human impacts, thus limiting the applicability of bioindicators in heterogeneous environments as Ceburnis and Valiulis report (1999). Accordingly, populations of indicator species may be influenced by factors other than the disturbance or stress (e.g., disease, parasitism, competition, predation), complicating our picture of the causal mechanisms of change. A second criticism of the use of bioindicators is that their indicator ability is scale-dependent. For example, a large vertebrate indicator (e.g., a fish) may fail to indicate the biodiversity of the local insect community. Thirdly, bioindicator species invariably have differing habitat requirements than other species in their ecosystem. Managing an ecosystem according to the habitat requirements of a particular bioindicator, may fail to protect rare species with different requirements. Finally, the overall objective of bioindicators is to use a single species, or a small group of species, to assess the quality of an environment and how it changes over time, and this can represent a gross simplification of a complex system.

Like all management tools, we must be conscious of its flaws. However, the limitations of bioindicators are clearly overshadowed by their benefits. Bioindicators can be used at a range of scales, from the cellular to the ecosystem level, to evaluate the health of a particular ecosystem. Bioindicators bring together information from the biological, physical, and chemical components of our world that manifest themselves as changes in individual fitness, population density, community composition, and ecosystem processes (Fernández et al., 2007). From a management perspective, bioindicators inform our actions what is and what is not biologically sustainable. Without the moss in the tundra, the cutthroat in the mountain stream, and the canary in the coal mine, we may not recognize the impact of our disturbances before it is too late to do anything to prevent them.

1.2. Mosses as Airborne Pollution Bioindicator (Challenges or Facts)

Bryophytes are green land plants which lack a vascular system and are simple both morphologically and anatomically. The growth potential in bryophytes is not as highly polarized as in vascular plants. Bryophytes grow in a variety of habitats especially in moist places on soil, rocks, trunks and branches of trees and fallen logs. They obtain nutrients directly from substances dissolved in ambient moisture. Some substances are probably absorbed directly from the substrate by diffusion through the cells of the gametophyte. Bryophytes are used as reliable indicators of air pollution (Fernández et al., 2007). They are exploited as *bryometers* instrument for measuring phytotoxic air pollution. Bryophytes, independently or together with lichens can be valuable organisms in developing an Index of Atmospheric Purity (IAP) which is based on the number, frequency-coverage and resistance factor of species (Ceburnis and Valiulis 1999). This index can provide a fair picture of the long-range effects of pollution in a given area (Markert et al., 2003). There are two categories of bryophytes in response to pollution:

- Bryophytes that are very sensitive to pollution and show visible symptoms of injury even in the presence of minute quantities of pollutants. These serve as good indicators of the degree of pollution and also of the nature of pollutant.
- Bryophytes that have the capacity to absorb and retain pollutants in quantities much higher than those absorbed by other plant groups growing in the same habitat. These plants trap and prevent recycling of such pollutants in the ecosystem in different periods of time. Analysis of such plants gives a fair idea about the degree of metal pollution.

Bryophytes are able to concentrate heavy metals in large amounts, greatly surpassing the absorbing capacity of vascular plants. The gametophytes of moss can accumulate iron 5-10 times more readily than the vascular plants. The concentration of Al, Ba, Cr, Cu, Fe, Ga, Ni, Pb, Ag, Ti, V, Zn and Zr is higher in bryophytes than the corresponding in angiosperms. The bryophytes are able to concentrate rare earth elements. Elements which are rarely found in other plants were found in bryophytes (Zechmeister et al., 2003; Balabanova et al., 2010; Barandovski 2012, 2013).

Metals are non-degradable and once released into the environment they become an integral part of the habitat. Bryophytes are able to concentrate heavy metals in larger amounts than those of vascular plants. Heavy metals are absorbed either from the atmosphere, or from the substrate, or from both sources. The older tissues of the plant have higher concentrations of the metallic ions compared to the younger portions (Fernández et al., 2012). The ability of mosses to accumulate heavy metals depends on the total leaf surface and the number of thin-walled parenchyma cells. Carpet forming bryophytes has proved to be rapid and inexpensive method for surveying heavy metal deposition in the terrestrial ecosystem (Ruhling and Tyler, 1968). The concentration of airborne material decreases in the plant tissues with a distance from the source of pollution and there is a wide variation in metal accumulation from species to species and from habitat to habitat under different microclimate conditions (Žibret and Šajn, 2008a). Among mosses the profusely branched and ramifying pleurocarps and the densely packed acrocarps are more efficient entrappers and absorbers of metal particles than the unbranched and erect acrocarps.

Mosses have been frequently used to monitor time-integrated bulk deposition of metals as a combination of wet, cloud, and dry deposition, thus eliminating some of the complications of precipitation analysis due to the heterogeneity of precipitation (Markert et al., 2003). Ectohydric mosses in particular draw negligible amounts of water and minerals from the soil, and almost entirely depend on atmospheric inputs of nutrients (Ruhling and Tyler, 1968). Because mosses have a high cation exchange capacity (CEC), they act as hyperaccumulators of metals and metal complexes. The metals are bound to the tissue with minimal translocation within the plant due to a lack of vascular tissue (Ruhling and Tyler, 1968). This results in biological tissue that can be analyzed to reveal time-integrated deposition (Zechmeister, 2003). Additional advantages of using mosses as heavy metal biomonitors include their stationary nature, widespread geographic distribution, and low genetic variability between populations. It has been shown that there is some experimental error due to heterogeneity in morphological characteristics and microenvironments among different populations (Zechmeister et al., 2003). There is also an incomplete understanding of the degree of mineral uptake by ectohydric mosses in direct contact with substrate (Gjengedal and Steinnes, 1990). Despite the accuracy and precision of precipitation analysis techniques, however, mosses offer an efficient, low-cost complement for determining metal concentrations at a large number of locations and offer analyses of biologically relevant fluxes at multiple scales.

Harmens and his European colleagues have found that mosses are reliable indicators of air pollution risks to ecosystems, because they get most of their nutrients direct from the air and rain, rather than the soil. Since 2000 the European moss survey has been conducted by a special international programme (ICP Vegetation). Moss data provides a better geographic coverage than measured deposition data and reveals more about actual atmospheric pollution at a local level (http://icpvegetation.ceh.ac.uk/).

1.3. Study Issue

Monitoring of air pollution has proved as the most useful technique for determining deposition of heavy metals and together with it, atmospheric pollution in different geographical areas. The Republic of Macedonia does not deviate from the global framework of air pollution with heavy metals. The results obtained from previous studies of air pollution suggest that the situation in the Republic of Macedonia is not very favorable. The main emission sources appear to be power plants, mines, as well as flotation of lead, zinc, copper and metallurgical plants for production of nickel, steel, lead, zinc and different ferroalloys (Barandovski et al., 2008; 2013 Stafilov et al., 2008, 2010, 2010a, 2010b; Balabanova et al., 2010, 2011, 2012, 2013; Bačeva et al., 2009, 2011, 2012, 2013).

In this study the total content of 23 elements was determined: Ag, Al, As, B, Ba, Ca, Cd, Co, Cr, Cu, Ga, Fe, K, Li, Mg, Mn, Mo, Na, Ni, Pb, Sr, V, and Zn in moss samples, taken from the copper mine and flotation Bučim near Radoviš (case 1); moss samples from Sasa zinc and lead mine and flotation plant (case 2); moss samples from Zletovo zinc and lead mine and flotation plant and flotation tailings dump near Probištip (case 3) moss samples from the ferronickel smelter plant in the region of Kavadarci (case 4) moss samples from the abandoned As-Sb-Tl mine in the Allchar area (case 5). Analyses were performed by application of atomic emission spectrometry with inductively coupled plasma (ICP-AES). However, for the Kavadarci region moss samples were analyzed by application of mass spectrometry with inductively coupled plasma (ICP-MS) as well determining the content of: Ag, Al, As, Ba, Be, Bi, Ca, Cd, Ce, Co, Cr, Cs, Cu, Dy, Eu, Fe, Ga, Gd, Ge, Hf, Hg, Ho, I, K, La, Li, Lu, Mg, Mn, Mo, Na, Nb, Ni, Pb, Rb, Sb, Sm, Sr, Tb, Th, Ti, U, V, Yb, Zn and Zr.

The obtained values for the content of respective elements in the investigated samples were statistically processed using statistical software for bivariate analysis, which showed the correlation of the content of chemical elements in samples; at the same time, multivariate statistic method was used to reveal the associations of the chemical elements.

Distribution maps for each element have been created separately based on the element's content. The statistically processed data and the distribution maps of individual elements make it possible to identify narrow areas with their highest concentration. In this way, the impact of the mines operation on the increase of the content of heavy metals in the air and the direction of the elements distribution in the area around the towns of Radoviš, Makedonska Kamenica, Probištip, Kavadarci and the Allchar region can be determined. This is of crucial importance having in mind the direct human exposure to these pollutants and their adverse effects on human health.

2. Investigated Areas

The Republic of Macedonia was included in the UNECE ICP Vegetation – Heavy Metals in European Mosses for the first time in 2002 (survey 2000/2001) and then again in 2005 and 2010 when atmospheric deposition of trace elements was studied over the entire territory of the country using samples of the terrestrial mosses *Hypnum cupressiforme* and *Homalothecium lutescens* given as *Camptothecium lutescens* in the previous published papers (Barandovski et al., 2012, 2013). Analyzing the results of the first survey, the most important emission sources were determined (mines and drainage systems and smelters near the towns of Veles, Tetovo, Kavadarci and Radoviš). A comparison of the results was made with the results obtained from similar studies performed in neighboring countries, as well as a comparison with more pristine territories in other parts of Europe (Barandovski et al., 2013). The results of the second survey showed increasing trends of elemental content in mosses that are connected with anthropogenic sources, such as Cd, Co, Pb, Hg and Ni in 2005 compared with the previous survey, but also a decreasing trend in the content of some elements in mosses such as As, Cr, Cu, Sb and Se (Barandovski et al., 2012).

The present study was conducted at four mines (three in function and one abandoned and one smelter plant location in the Republic of Macedonia. In the eastern part of the country the appearance of certain metals (Cu, Au, Mg, Al, Sc, Ti, V) in the air is related to the presence of the copper mine and flotation plant Bučim, near the town of Radoviš (Stafilov et al., 2010b; Balabanova et al., 2010; Balabanova et al., 2011). This was one of the investigation areas influenced also by the former iron open pit mine, Damjan (Serafimovski et al., 2006). The second case study refers to the mine and flotation plant Sasa located in the central eastern part of the country where higher contents of Pb, Zn and Cd were detected (Barandovski et al., 20013). Zletovo lead-zinc mine and Probištip flotation plant have a total of ~400 km^2 area located in the eastern part of the Republic of Macedonia. The fourth study area is located in the south-central part of the country, where the appearance of some metals (Co, Cr, Fe and Ni) in the air is related to the presence of the ferronickel smelter plant near the city of Kavadarci (Barandovski et al., 2008, 2012, 2013; Stafilov et al., 2008, 2010; Bačeva et al., 2009; 2011). Near the fourth case study area is located the fifth case study area, which is around the abandoned As-Sb-Tl mine at the locality of Allchar (Bačeva et al., 2013). As a result of these anthropogenic activities, increased distribution of certain heavy metals in the air and their deposition in the environment were expected.

2.1. Bučim Copper Mine Case Study

The first case study area is located in the eastern part of the Republic of Macedonia with a surface area of 20 km (W–E) × 20 km (S–N), a total of 400 km^2, limited by the coordinates N: 41°32' – 41°44' and E: 22°15' – 22°30' (Figure 1). The region is characterised by a moderate continental climate (Lazarevski, 1993). The altitude varies between 350 and 1000 m. The average annual temperature is around 10°C. The average annual rainfall amount is 563 mm with large variations from year to year. The most frequent winds in the region are those from the west with a frequency of 199 ‰ and speed of 2.7 m s^{-1}, and winds from the east with a frequency of 124 ‰ and speed of 2.0 m s^{-1}.

One of the major emission sources of certain metals in the eastern part of the R. Macedonia is the copper mine and flotation Bučim, near the town of Radoviš. The mine and ore processing plant have been in function from the late seventies of the last century. Ore excavation is from an open pit and the ore tailings are stored on the open, in the mine vicinity. The produced copper ore from the mine is processed in the flotation plant; after the flotation of copper minerals, the flotation tailings are separated, disposed of and deposited on a dump site in an adjacent valley near the village Topolnica. The copper mine Bučim is located in the north-west part of the studied area. The Bučim mine territorially and administratively belongs to the municipality of Radoviš, and is located about 14 km from the town. In the close vicinity of the mine there are two settlements, the villages Bučim and Topolnica. Mine activities cover 7 km^2 of total mine surfaces, 4 km^2 for the placement of ore tailings and the rest of the land belongs to the open ore pit and to the ore processing plant. The main ore contents are the following: 0.3 % Cu, 0.3 g t^{-1} Au, 1 g t^{-1} Ag, 13 g t^{-1} Mo, and 1–4 % pyrite; the igneous rocks have been altered to clays and micas. The important metallic minerals found are chalcopyrite, pyrite, and bornite, with small amounts of galena, sphalerite, magnetite and hematite (Stafilov et al., 2010b; Balabanova et al., 2011).

The Bučim mine and the ore processing plant have been operating since 1979 and it is assumed that the mine has about 40 million tons of ore reserves. Ore tailings are dropped out by the dampers from the open ore pit at an open site near the mine. The ore tailings deposit occupies a surface of 0.80 km^2, located southwest of the open ore pit, near the regional road Štip-Strumica. The ore tailings deposit contains about 150 million tons of ore tailings. Exposure of ore tailings to constant air flow and wind leads to distribution of fine dust in the air (Stafilov et al., 2010b).

The flotation plant produces 4 million tons of copper ore annually. In the process of flotation of copper minerals, the average annual amount of produced flotation tailings is approximately 3.95 million tons. These tailings are drained and disposed of on a dump near the mine. The dump is located to the east of the flotation plant, at a distance of about 2.2 km.

2.1.1. Geological Description

The investigated area represents a part of the Vardar structural zone, separated from the other structural zones during the Caledonian, and subjected to strong tectonic processes during the Herzynian orogenesis. The structural relations were further complicated by the Alpine orogenesis (Rakićević et al., 1969).

The study area is characterised by the following main geotectonic structural units: (1) the Kriva Lakavica basin (2) the Smrdeš-Gabreš syncline, (3) the Radoviš basin, (4) the Radoviš anticline divided to (4a) the Štip Block and (4b) the Bučim Block (Figure 2). The Radoviš anticline represents the eastern boundary of the Vardar zone towards the Serbian-Macedonian mass. These two large structural units are separated by a deep NW-SE fault (Hristov et al., 1965). The Bučim-Damjan-Borov Dol area is divided into two tectonic blocks. The Bučim tectonic block and the southern tectonic block Damjan are parts of the Vardar zone. The blocks are divided by a fault of first order in the SE direction. Despite the disposition in two different tectonic blocks, the metallogenic area is unified, based on the similarities of Tertiary magmatism and the analogous ore mineralisation. The Bučim copper-porphyry deposit with additional gold mineralisation is found in the northern block (Stefanova et al., 2004).

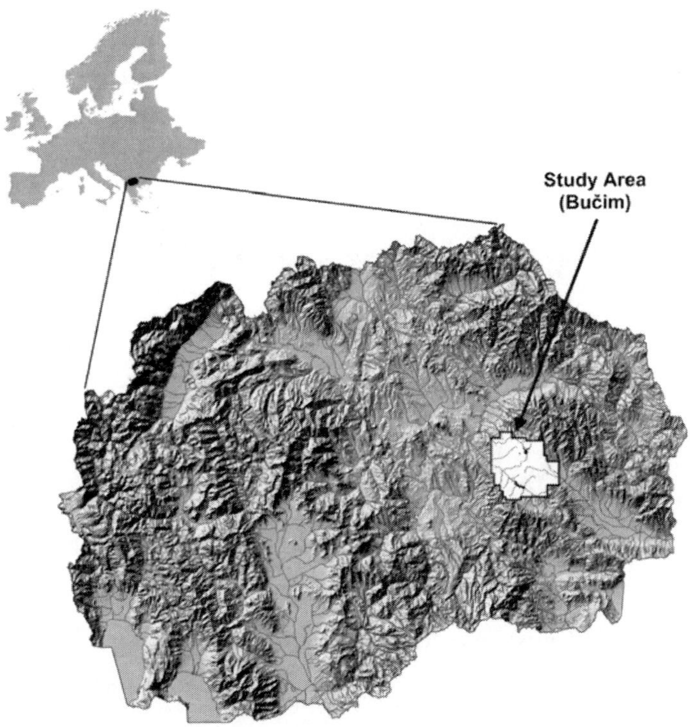

Figure 1. Location of the study area in the Bučim" copper mine region, Republic of Macedonia (Balabanova et al., 2010).

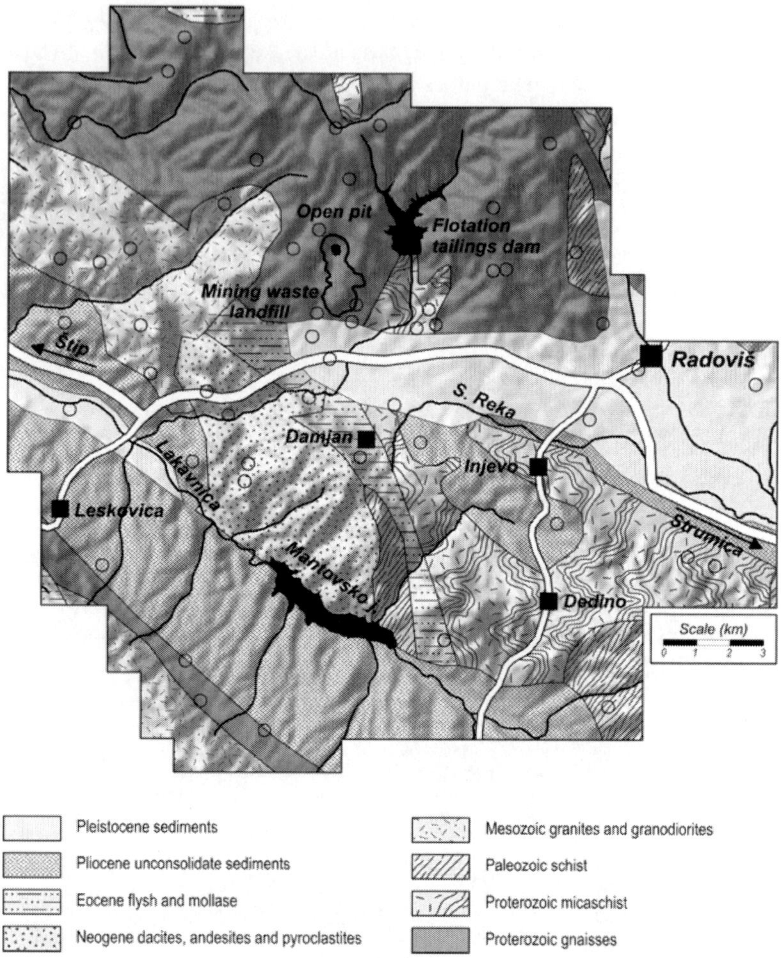

Figure 2. Geological map of Bučim copper mine, with the moss sampling locations (Balabanova et al., 2010).

2.2. Sasa Lead-Zinc Mine and Flotation

A total of 300 km^2 area was monitored, limited with coordinates N: 41°32' – 41°44' and E: 22°15' – 22°30', located in the eastern part of the Republic of Macedonia (Figure 3). The Sasa lead-zinc mine environment was monitored as potentially polluted area with anthropogenic introduction of high contents of certain heavy metals. The region is characterized by moderate continental climate. The altitude varies between 400 and 1500 m. The average annual rainfall amounts to 563 mm with large variations from year to year. Most frequent winds in the region are those from the west with frequency of 199 ‰ and 2.7 m s^{-1} speed, and winds from the east with frequency of 124 ‰ and 2.0 m s^{-1} speed. The climatic conditions in the region allow air-distribution of fine dust particles generated as a result of mining activities and exposure of flotation tailings at open. The Sasa mine is located in the north-west part of the study area. The pollution source occupies an area of about 80 km^2. The Sasa mine has been in operation for over 45 years, annually yielding around 90,000 tons of Pb-Zn high quality concentrate.

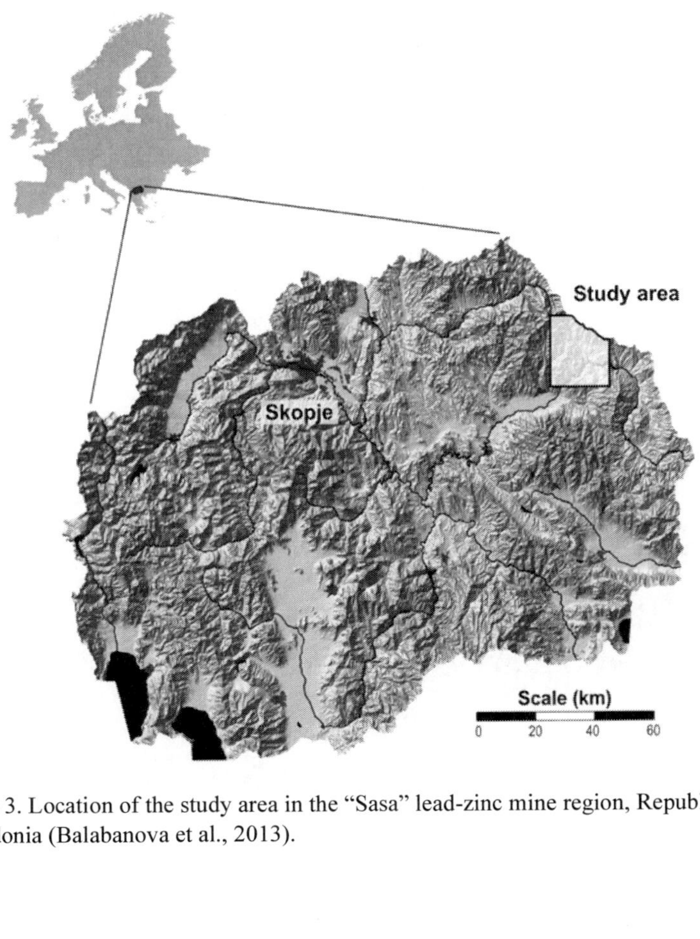

Figure 3. Location of the study area in the "Sasa" lead-zinc mine region, Republic of Macedonia (Balabanova et al., 2013).

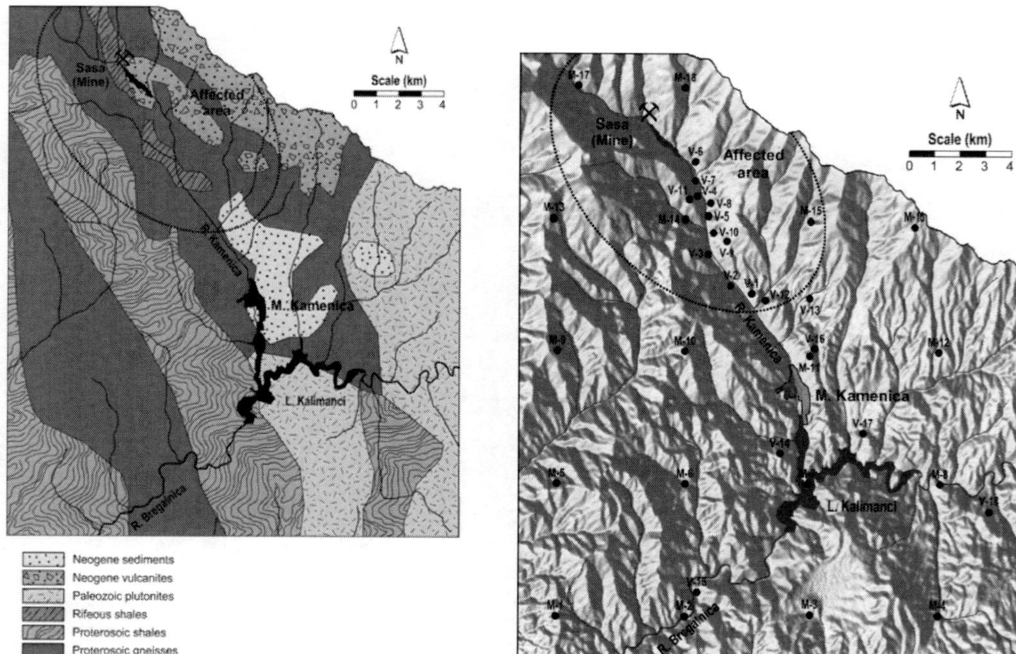

Figure 4. Generalized geology of the “Sasa” lead-zinc mine region (left) with the moss sampling locations (right) (Balabanova et al., 2013).

2.2.1. Geological Description

The Sasa Pb-Zn deposit lies within the Sasa-Toranica mining district in the Osogovo Mountains in the eastern part of Macedonia (Figure 4). The geology of the Toranica-Sasa ore field comprises various rocks of both metamorphic and igneous origin, with the latter of Tertiary age. The most economically valuable mineralisation is closely related to the quartz-graphite schists, with the ore consisting mainly of: galena, sphalerite, chalcopyrite and pyrite. Further studies have revealed more details about the complexity of the deposit's mineralogy, including the presence of: galena, sphalerite, chalcopyrite, pyrite, pyrrhotite, magnetite, martite, bornite, enargite, tetrahedrite, marcasite, barite, native gold, cubanite and native bismuth (Serafimovski et al., 2006). In 2003 a major environmental disaster took place in the investigated region when part of the Sasa mine tailings dam collapsed and caused an intensive flow of tailings material through the Kamenica valley. Between 70,000 and 100,000 m^3 of tailings material was discharged into Lake Kalimanci, causing significant ecological damage and occurrence of potential hazardous conditions in relation to air pollution (Vrhovnik et al., 2011).

2.3. Zletovo Lead-Zinc Mine and Probištip Flotation Plant

A total of ~400 km^2 area was monitored, limited with coordinates N: 41°51'–42°08' and E: 22°01' – 22°25', located in the eastern part of the Republic of Macedonia (Figure 5). The lead-zinc mine and flotation environment was monitored as potentially polluted area with anthropogenic introduction of high contents of certain heavy metals. In terms of climate, the field is located in the south-north temperate zone, including areas which are experiencing the effects of the Mediterranean climate (Kočani Valley and terrain) and Osogovo Mountain where mountainous climate dominates. This geographical position which conditions the climate is characterized by elements of moderate continental climate (Lazarevski, 1993). The altitude varies between 300 -1500 m. The average annual rainfall amounts to about 600-650 mm with large variations from year to year. Most frequent winds in the region are those blowing from the west with frequency of 199 ‰ and 2.7 m s^{-1} speed, and winds from the east with frequency of 124 ‰ and 2.0 m s^{-1} speed. Climatic conditions in the region allow air-distribution of fine dust particles generated as a result of mining activities and exposure of flotation tailings at open.

(a)

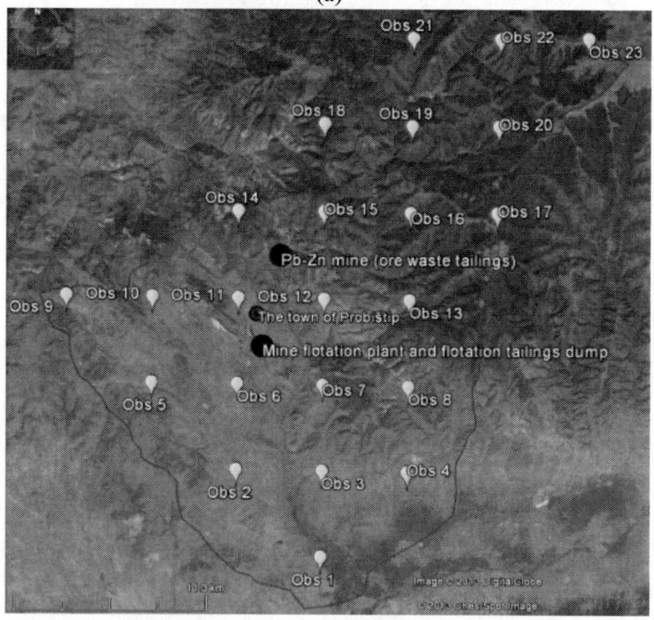

(b)

Figure 5. (a) Location of the study area in the Zletovo lead-zinc mine region, Republic of Macedonia, (b) with the moss sampling locations.

2.3.1. Geological Description

The Zletovo Pb–Zn deposit is situated along the active continental margin and the geology is associated with the Tertiary volcanism and the hydrothermal activity of the area. The Zletovo mine is located about 5 km NW from the village of Zletovo and about 7 km from the town of Probištip (Figure 5). The continuous exploitation of the mine started after the Second World War. It has annual capacity of 300,000 tons (9% Pb and 2% Zn) and significant concentrations of Ag, Bi, Cd, and Cu. The mine has been active until this day with production of Pb–Zn concentrate. Mineral association comprises galena (principal ore mineral) and sphalerite, with subordinate pyrite, lesser amounts of siderite, chalcopyrite, and occasional pyrrhotine, marcasite, and magnetite. Ore is concentrated by flotation at Probištip and the tailings are stored in two impoundments situated in the adjacent valleys (Serafimovski et al., 2006).

2.4. Ferronickel Smelter Plant Case Study (Fe-Ni Industry)

In the south-central part of the country the appearance of certain metals (Co, Cr, Fe and Ni) in the air is related to the presence of the ferronickel smelter plant near the town of Kavadarci, (Barandovski et al., 2006; 2008; Stafilov et al., 2008, 2010; Bačeva et al., 2009; 2011). This smelter plant uses nickel ore from the Ržanovo mine, located about 30 km south of the plant. In the last several years apart from the ore from the Ržanovo mine, nickel ore originating mainly from Indonesia and some other countries (2-2.5 %) has been used. It is well known that nickel ores also contain cobalt (average of 0.05 % for Ržanovo ore) (Maksimović, 1982; Boev and Jankovic, 1996). Therefore, the dust from this plant has similar content with the ore used as a raw material, including some of the heavy metals like nickel, cobalt and chromium. For that reason, the goal of this work was to determine the total deposition (deposited dust) in the atmosphere in Kavadarci and its surroundings.

The investigated area (Figure 6) is located in the south-central part of Macedonia, covering approximate area of about 600 km^2 [20 km (W-E) x 30 km (S-N)], with limits at coordinates N: 41◦10_ – 41◦30_ and E: 21°52_– 22°09_. The town Kavadarci is located in the Tikveš valley, about 100 km south of the capital Skopje. It is the main vine production region in Macedonia. The municipality of Kavadarci (38,741 inhabitants; 992.44 km^2) is composed of the town of Kavadarci (28,000 inhabitants) and 39 settlements.

The urban area is located at an altitude of 200–300 m and is surrounded by hills from the east and south sides of the valley with height difference approximately between 300 and 770 m. The west side of the valley is surrounded by Ljubaš and Dabov Vrv hills and mountain Klepa. The east side of the valley is surrounded by mountain Serta, and the south side by Vitacevo plateau and mountain Kožuf. The ferronickel smelter plant is located in the west part of the study area (Figure 6). It is a two-line, rotary kiln electric furnace facility with the biggest rectangular electric furnaces of their kind in the world. The plant has been in operation since 1982. At the time of its acquisition the annual nickel production was approximately 5,000 tons. However, after the implementation of technical additions, the plant is now fully operational and the production is steadily increasing to 20.000 t/y. The nickel ore processed in the factory comes by conveyor from the open-pit Ržanovo mine, located close to the production facility. In order to optimise the production, the Ržanovo ore is then blended with higher-grade ores imported from Indonesia, The Philippines, Greece, Turkey and Albania. The content of Ni ranges from 1 to 2.5% Ni. On each line, the plant operates a rotary dryer that feeds into a rotary kiln where the nickel ore is dried and reduced and then fed into an electric furnace for complete reduction into a molten ferronickel alloy. The ferronickel alloy is then placed into an oxygen converter and blown with oxygen to reduce the iron content, thereby increasing the nickel content.

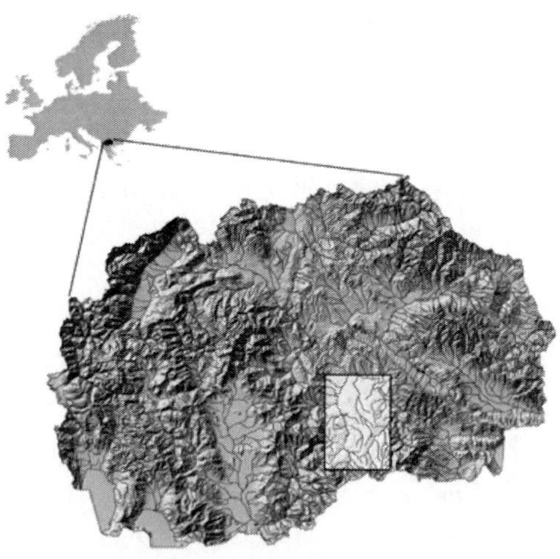

Figure 6. Location of the study area in the region of "FENI industry" smelter plant, Republic of Macedonia (Bačeva et al., 2012).

2.4.1. Geological Description

The geological description of the investigated area is presented in the Geochemical Atlas of Kavadarci and the Environs (Stafilov et al., 2008, 2010). The oldest formations have direction NW-SE and belong to the inner parts of the Vardar zone. The Lower Paleozoic (Pz) metamorphic complex is present with two series: amphibole and amphibole chlorite schist with marbles and phylite layers. Serpentinite is present in the form of narrow belts along the ruptures inside the Vardar zone. The uttermost part in the SW of the study zone is covered with marbles and dolomites probably from Devonian ages. Over the Paleozoic, Mesozoic (Mz) formations have been developed, mainly from Late Cretaceous ages. Paleozoic and Mesozoic rocks cover approximately 39 km^2 in the SW part of the investigated area.

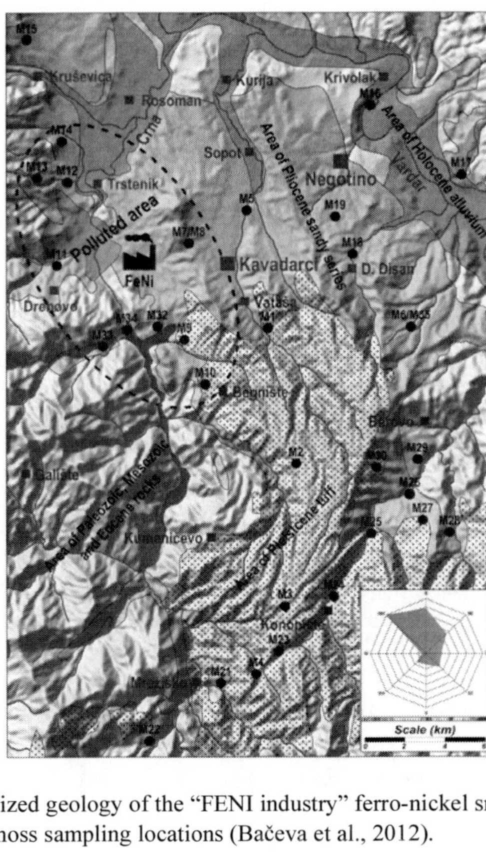

Figure 7. Generalized geology of the “FENI industry” ferro-nickel smelter plant region, with the moss sampling locations (Bačeva et al., 2012).

Complexes of Tertiary and Quaternary sediments cover the most of the study area. The Upper Eocene ($^4E_3$) flysch sediments and yellow sandstones are developed along Vardar, Crna Reka and Luda Mara valleys and the marginal part of the Tikveš basin. Those sediments with depth of 3500 m cover approximately 34 km^2 mainly in the Northern part of the investigated area. The Pliocene sediments fill the Tikveš basin, limited with the Vardar on the North, and Paleozoic-Mesozoic formations with direction NW-SE. This sequence is represented mainly by sandy series. Pliocene (Pl) sediments cover the biggest part (about 182 km^2) in the central part of the investigated area. SE from Kavadarci the Quaternary (Q) pyroclastic vulcanites are found. They are presented by tuffs, breccias and agglomerates, and cover approximately 25 km^2.

2.5. Allchar As-Sb-Tl Mine Case Study

The Kožuf area is a large volcanic complex situated in the south of the Republic of Macedonia. The Allchar As-Sb-Tl locality is comprised of hydrothermal volcanogenic deposits situated in the NW of the Kožuf Mountain, close to the border between the Republic of Macedonia and Greece (Figure 8). From the geotectonic point of view, ore mineralisation is related to the Pliocene volcano-intrusive complex located between the rigid Pellagonian block in the west, and the labile Vardar zone in the east characterized by ring-radial structures (Boev and Jelenković 2012). From the metallogenic point of view, the Allchar deposit belongs to the Kožuf ore district as part of the Serbo-Macedonian metallogenetic region.

2.5.1. Geological Description

The Allchar deposit has a complex mineral composition comprised of sandstone and claystone, followed by bedded and massive carbonate rocks of the Middle and Upper Triassic, representing the deposit bedrock. Quartz-sericite-feldspar schists are developed along the eastern flank of the deposit, while the central part is built of dolomite, marble, and sporadically limestone. The dolomite series lie under marble while the Mesozoic rocks are unconformably overlaid by Pliocene cover and glacial till. Dolomite unconformably overlies the Mesozoic bedrock, particularly in the central, northern and southwest parts of the deposit. This unit of volcanic sedimentary provenance is commonly mineralized. The Pliocene felsic-tuffs unit covers a large portion of the Allchar deposit. This volcanic sequence includes ash, crystal tuffs, tuff breccias and lacustrine tuffaceous sediments. The tuffs deposited in the sublacustrine basins in the southern part of Allchar show bedding and contain tuffaceous sedimentary clay material (a volcano-sedimentary series) as given in Figure 9.

The Allchar deposit comprises several ore bodies within a zone 2 km long and around 300-500 m wide. Mineralisation was associated with hydrothermally altered wallrocks including the Triassic carbonates (dolomites and marbles), the Tertiary magmatic rocks and a volcano-sedimentary sequence (tuffaceous dolomite). Silicification and argillitization were the most predominant alteration products, and quartz was very abundant in hydrothermally altered volcaniclastites. The alteration was generally believed to be associated with Plio-/Pleistocene andesite volcanism and latite intrusion, which extends from mountain Kožuf in the Republic of Macedonia to mountain Voras in Greece (Jelenković and Boev, 2011; Jelenković et al., 2011; Boev and Jelenković, 2012).

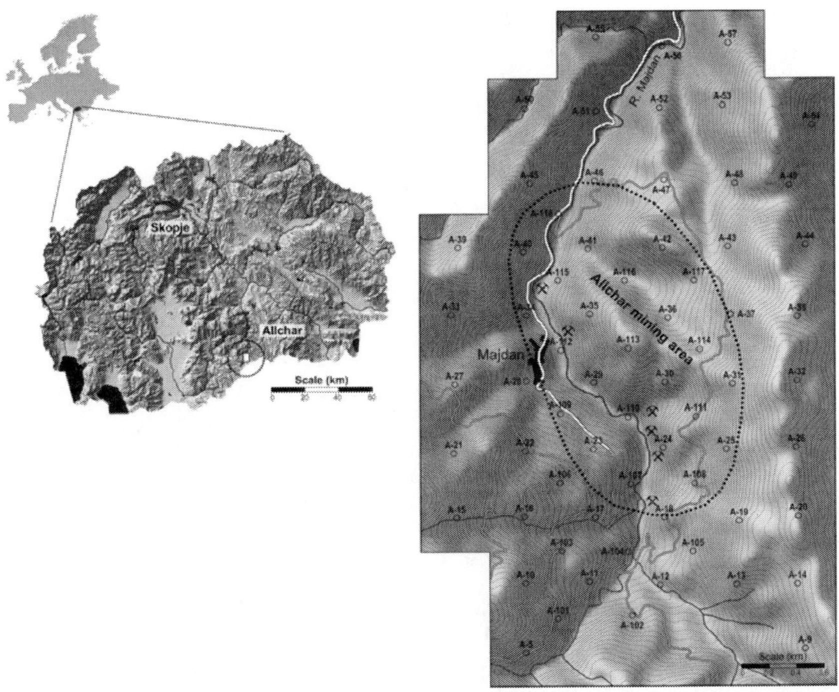

Figure 8. Location of the study area in the Allchar As-Sb-Tl mine, Republic of Macedonia (a), with the moss sampling locations (b), (Bačeva et al., 2013).

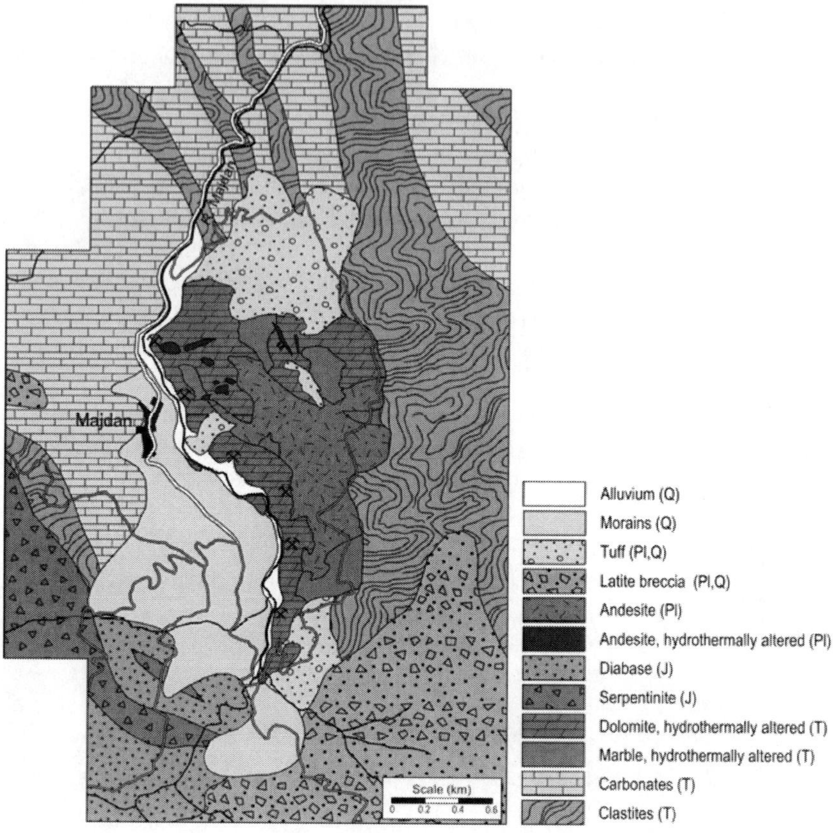

Figure 9. Generalized geology of the Allchar As-Sb-Tl mine region (Bačeva et al., 2013).

3. Materials and Methods

3.1. Moss Sampling

Moss sampling was performed according to the guidelines set out in the experimental protocol of the 2005/6 survey (ICP Vegetation, 2005). Each sampling site was located at least 300 m from main roads and populated areas and at least 100 m from any road or single house. In forests or plantations, samples were collected as far as possible in small open spaces to preclude any significant effect of canopy drip. Sampling and sample handling were carried out using plastic gloves and bags. Each sample was a composite of about five sub-samples. Dead material and litter were removed from the samples and only the last three years' growth of material was used for the analyses. Samples were refrigerated, deep-frozen or dried at room temperature and stored under adequate conditions until chemical analysis. Moss species *Hypnum cupressiforme* (Hedw.) and *Camptothecium lutescens* (Hedw.), *Homolothecium sericeum* (Hedw.), and *Scleropodium purum* (Hedw.) Limpr., were collected from the whole study area as most characteristic moss species for the flora of the Republic of Macedonia. In all investigated cases dominant moss species were *Hypnum cupressiforme* and *Camptothecium lutescens*, while the *Homolothecium sericeum* and *Scleropodium purum* were less findable species.

3.1.1. Moss Sampling in Bučim Cu Mine and Flotation Environ

The collection of moss samples was performed according to the protocol adopted within the European Heavy Metal Survey. Moss species were collected according to the previously defined sampling network around the copper mine Bučim in an area of 400 km^2 (Figure 2). Total of 52 moss samples were collected, including *Hypnum cupressiforme* as dominant species (72%), *Camptothecium lutescens* (22%) as less present species in the area and *Scleropodium purum* found at only 3 locations. Moss species were readily available and could be easily found at each designated location. The occurrence was probably due to the geographical location of the region and the predominant altitude over 400 m. The sampling locations of the collected moss species are given in Figure 2, with the generalized geology of the region.

3.1.2. Moss Sampling in Sasa Pb-Zn Mine Environ

Total of 36 moss samples of *Hypnum cupressiforme* (Hedw.) and *Camptothecium lutescens* (Hedw.) were collected from the whole case study area as characteristic moss species (Figure 4). Depending on the conditions and the accessibility of the locations, the species that are available and typical for the region were collected. Due to the geographical constitution of the ground conditions and pollution, two ways of collecting moss samples were applied. Random samples were collected all over the canyon where flotation tailings were leaking, and introduction of higher contents of Pb and Zn in its environment was expected (marked as V-1 to V-16). The second group of samples was collected according to the sampling network 5 x 5 km (marked as M-1 to M-18) as presented in Figure 4.

3.1.3. Moss Sampling in Zletovo Pb-Zn Mine Environ

Total of 23 sampling locations were defined in the investigated area (Figure 5). Three local characteristic moss species were used as biomonitors (*Hypnum cupressiforme, Scleropodium purum* and *Campthotecium lutescens*) reproducing a sampling network with density of 5x5 km. The dominant moss species among all collected species was *Hypnum cupressiforme* (64%). The moss sampling protocol was performed according to the set standard rules for collection of such samples.

3.1.4. Moss Sampling in Fe-Ni Smelter Plant Environ

Samples of three moss species *Hypnum cupressiforme*, *Campothecium lutescens* and *Homolothecium sericium* were collected at 31 locations (Figure 7), evenly distributed throughout the country. Most of the moss samples were collected around the ferronickel factory. Owing to the nature of the field, the sampling conditions were very specific. Based on this assumption, random sampling was applied. The altitude significantly varies over the region, and therefore the samples were collected at the discretion of the analyst and in accordance with the geographical conditions.

3.1.5. Moss Sampling in Allchar As-Sb-Tl Mine Environ

Samples of 8 different moss species [*Hypnum cupressiforme*, *Homalothecium lutescens* (Hedw.) Robins., *Lencodon scinroides* (Hedw.) Schwaegr., *Brachythecium salebrosum* (Web & Mohr.) B. S. & G., *Scleropodium purum* (Hedw.) Limpr., *Homalothecium sericium* (Hedw.) B. S. & G., *Brachythecium albicans* (Hedw.) B. S. & G., *Brachythecium glareosum* (Spruce) B. S. & G.] were collected during the summer 2011 at a total of 69 locations (Figure 8). Different moss species were taken from different sites according to their availability. Sampling was carried out according to the European moss survey protocols (Rühling and Steinnes, 1998).

3.2. Moss Species Treatment (Wet Digestion)

The moss sampling protocol was performed according to the set standard rules for collection of such samples, as mentioned above. In the laboratory, the samples were cleaned from extraneous plant material and air-dried at room temperature. Only the last three year's growth of moss material was used without washing for the analysis. Moss samples (0.5 g) were placed in Teflon digestion vessels, 5 mL concentrated nitric acid (HNO_3, 69%, trace pure for trace analysis) and 2 mL H_2O_2 (30%, *m/V* trace pure for trace analysis) were added, and the vessels were capped closed, tightened and placed in the rotor of the Mars microwave digestion system (Mars, CEM, USA). The moss samples were digested at 180°C. After cooling, the digested samples were quantitatively transferred into 25 mL calibrated flasks as previously given by Balabanova et al. (2010) and Bačeva et al. (2012).

3.3. Analysis of Total Elements Contents (AAS, ICP-AES and ICP-MS Techniques)

The investigated elements were analyzed by application of atomic emission spectrometry with inductively coupled plasma (AES-ICP), mass spectrometry with inductively coupled plasma (ICP-MS) as well as electrothermal atomic absorption spectrometry (ETAAS) and cold vapour AAS (CV-AAS). The following elements were analyzed: Ag, Al, As, Ba, Be, Bi, Ca, Cd, Ce, Co, Cr, Cs, Cu, Dy, Eu, Fe, Ga, Gd, Ge, Hf, Hg, Ho, I, K, La, Li, Lu, Mg, Mn, Mo, Na, Nb, Ni, Pb, Rb, Sb, Sm, Sr, Tb, Th, Ti, U, V, Yb, Zn and Zr.

Atomic emission spectrometer with inductively coupled plasma (Varian 715-ES) was used for determination of Al, Ca, Cr, Cu, Fe, K, Li, Mg, Mn, Na, Ni, Pb, Sr, V and Zn, while ETAAS for determination for As, Cd and Co (Varian SpectrAA-604Z). For AES-ICP instrument calibration and quantitative determination of each element's content, a commercial standard mix solution (11355-ICP Multi Element Standard IV, Merck) was used. The correlation coefficient of calibration curve for each element was 0.999. In order to check the possible background contamination, blank samples were used and processed simultaneously with field samples. The method detection limit was calculated based on average measuring of the blank sample ($\gamma \pm 3\delta$). For all laboratory samples and standard solutions, treated ultra pure water was used. The QC of the applied techniques was performed by standard addition method, and it was found that the recovery for the investigated elements ranges between 98.5–101.2 % for ICP-AES and between 96.9 % – 103.2 % for ETAAS.

For ICP-MS measurements a SCIEX Perkin Elmer Elan DRC II (Canada), inductively coupled plasma mass spectrometer (with quadruple and single detector setup) was used. A total of 46 elements (Ag, Al, As, Au, Ba, Be, Ca, Cd, Ce, Co, Cr, Cs, Cu, Dy, Er, Eu, Fe, Ga, Gd, Ge, Hg, Ho, K, La, Li, Lu, Mg, Mn, Mo, Na, Nb, Ni, P, Pb, Rb, Sb, Sm, Sr, Tb, Th, Ti, U, V, Yb, Zn, Zr) were determined by mass spectrometry with inductively coupled plasma (ICP-MS). The instrument's running parameters were checked and adjusted before every batch of measurements, using a solution with 1 μg ml^{-1} In, 1 μg ml^{-1} Ce, 10 μg ml^{-1} Ba and 1 μg ml^{-1} Th and Mg 1 μg ml^{-1}. Oxide levels and double ionized levels were kept under 3%, background for both low and high mass was under 1 cps and all the other parameters were chosen considering the best signal/noise ratio. The dynamic reaction chamber (DRC) was used in RF-only mode (no gas) and its parameters optimization have been optimised elsewhere (Tănăselia et al., 2008). For sample introduction system, a classic set-up was used, consisting of a peristaltic pump, a Meinhard nebuliser and a cyclonic spray chamber, where fine aerosols are formed that go directly into plasma. All other reagents were supplied by Merck. 18 MΩ cm^{-1} DI water was prepared in the laboratory, using a Millipore-Milli-Q® ultrapure water purification system.

All measurements were done using the quantitative method (Total-Quant) supplied by Elan 3.4 software that uses a response factor calibration curve which was obtained by calibration in multiple points, low, medium and high mass, for optimum set-up, using a multi-element Merck VI standard solution, diluted to mimic real sample composition. The drawback is that the accuracy tends to be worse than a proper quantitative method for some elements; however the main advantage is the large mass interval that can be studied (up to 65 elements per each sample during a single run), a good choice for screening type measurements that requires high throughput of samples with many elements of interest.

For this study, M2 and M3 moss samples certified reference materials (Steinnes et al., 1997) were used to check method accuracy for all considered elements, and the difference between measured and certified values was within 15%.

The theoretical limit for ICP-MS methods are in ppt (ng/L) range for the majority of the elements. Matrix effects above 1 ppb (μg/L) threshold while using Total-Quant were not observed during our study. For some elements, values between these two levels were further investigated using more complex quantitative methods.

Optimal conditions for the analysis of arsenic and cobalt are provided by ETAAS and of mercury by CV-AAS as described by Balabanova et al. (2010).

3.4. Data Processing

The obtained values for the contents of the investigated elements were statistically processed using basic descriptive statistics. Bivariate statistical approach was used to check data about elements contents correlations. For this reason, the linear coefficient of correlation was used. Two-dimensional scatter plots were used to visualize the relations between two data sets. Individual data points were represented by point markers in two-dimensional space, where axes represent the variables.

3.2.1. Data Transformations

Data distribution was examined by application of normality tests. Transforming data means performing the same mathematical operation on each piece of original data. If the original data is simply multiplied or divided by a specific coefficient or a constant is subtracted or added we talk about linear transformations. But these linear transformations do not change the shape of the data distribution and, therefore, do not help to make data look more normal.

It is often observed that environmental variables are Log-normal or positively skewed (Zhang and Selinus, 1998), and data transformation is necessary to normalize such data sets. Logarithmic transformation is widely applied in order to normalize positively skewed data sets. However, it is observed that data sets in environmental sciences do not always follow the Log-normal distribution (Zhang and Zhang, 1996). In such cases, a power transformation is needed, and Box–Cox transformation is one of the most frequently used of these (Box and Cox, 1964; Zhang and Zhang, 1996). Each data transformation can dampen the difference between extreme values. The Box-Cox transformation is given by the following Equation (1) and Equation (2):

$$y = \frac{x^{\lambda} - 1}{\lambda}; \lambda \neq 0 \tag{1}$$

$$y = \ln(\lambda); \lambda = 0 \tag{2}$$

where y is the transformed value, and x is the value that is to be transformed. For a given data set (x_1, x_2 … x_n), the parameter λ is estimated based on the assumption that the transformed values (y_1, y_2 … y_n) are normally distributed. When λ=0, the transformation becomes the logarithmic transformation.

3.3.2. Multivariate Statistical Analysis

Multivariate statistical methods (cluster and R-mode factor analyses) were used to reveal the associations of the chemical elements. The factor analysis was performed on variables standardized to zero value and unit standard deviation (Filzmoser et al., 2005; Žibret and Šajn, 2010). For measuring similarity between variables, the product-moment correlation coefficient (r) was applied. There are various rotational strategies that have been proposed (Žibret and Šajn, 2010).

The goal of all of these strategies is to obtain a clear pattern of loadings, that is, factors that are somehow clearly marked by high loadings for some variables and low loadings for others.

The elements with low communalities were excluded because of their lack of significant associations. In this study, the varimax method was used for orthogonal rotation. As it was said before, we intend to find a rotation that maximizes the variance on the new axes, or put another way, we want to obtain a pattern of loadings on each factor that is as diverse as possible, lending itself to easier interpretation.

The Cluster Analysis module was used for computing various types of distance measures, or the user can compute a matrix of distances himself. The purpose of this algorithm is to join together variables (elements, in this case) into successively larger clusters, using some measure of similarity or distance. These distances can be based on a single dimension or multiple dimensions. The most straight forward way of computing distances between objects in a multi-dimensional space is to compute Euclidean distances. This is probably the most commonly chosen type of distance.

The dendrograms were performed using the linkage distance reported as *D*link/*D*max (quotient between the linkage distances for a particular case divided by the maximal linkage distance), similar presented by Dolegowska et al. (2013).

3.3.3. Spatial Distribution

The universal kriging method with linear variogram interpolation was applied for the construction of spatial distribution maps of each factor score (Žibret and Šajn, 2008b). The universal kriging, the optimal linear unbiased predictor, considering linear trend with altitude and longitude was used for MYP spatial interpolation.

Seven classes of the following percentile values were selected: 0–10, 10–25, 25–40, 40–60, 60–75, 75–90 and 90–100. The cross-validation technique was used for the evaluation of the kriging results. The influential surrounding is the area around the prediction location, from which the measured values are taken into account by kriging. The quality of the kriging predictions was examined on the basis of cross-validation results and on the basis of a kriging variances analysis.

4. RESULTS AND DISCUSSION

4.1. "Bučim" Copper Mine Case Study (Case 1)

A total of 16 elements were analyzed by application of atomic emission spectrometry with inductively coupled plasma. The descriptive statistics of the analyzed elements is shown in Table 1. Values of Al, Ca, Fe, and K are given in % and the remaining elements in mg kg^{-1}. On the basis of normality tests compared with histograms of distribution of the content of all analyzed elements in moss, normality was assumed for natural values only for Ba; for the rest of the elements, normality was established on the basis of the logarithms of their contents. Lithogenic elements (Al, As, Cr, Fe, Mn, Ni) deposit in the environment corresponding to the geology of the region, and natural enrichment was not assumed. Aluminum content in moss tissue ranges from 0.47-8.51% and compared with the values for the Al content in moss on the whole territory of the R. Macedonia (0.054-0.87%), certain degree of enrichment was detected. However, the lower median value of 1720 mg kg^{-1} compared with 1900 mg kg^{-1} shows stability and non-significant anthropogenic influence on the Al content in the Cu mine environ. Arsenic ranges from 0.14-13.7 mg kg^{-1}, with the median value of 1.55 mg kg^{-1}. The maximum value was found very close to the mines which signalizes anthropogenic introduction of As due to the wind distribution of ore waste and flotation tailings.

Chromium as lithogenic element naturally deposits in areas with parts of Pleistocene sediments and Neogene dacites, andesites and pyroclastites. Manganese content varies in different sites within the study area from 59-439 mg kg^{-1}, with the median value of 144 mg kg^{-1}, which, compared with the value of 130 mg kg^{-1} from the 2010 research on the whole territory of the R. Macedonia , indicates nonspecific variation.

Nickel can be considered as lithogenic element but also as an essential element of the moss plant tissue. The Ni content varies in range from 2-30 mg kg^{-1}, with no significant deviation from the Ni content in moss tissues on the whole territory of the R. Macedonia (1.3-50 mg kg^{-1}).

The presence of biogenic elements (Ba, Ca, K, Na, Sr) was determined ranging from micro to macro content. Barium and strontium contents are most frequently correlated in their deposition in the environment. Their content varies from 11-66 mg kg^{-1}, and 13-55 mg kg^{-1} respectively. These values do not significantly vary from the correspondent values given by Barandovski et al. (2013).

A significant variation in these elements content was not determined, compared with the values of the contents of Ca, K and Na in moss on the whole territory of the R. Macedonia. Variation in moss biogenic elements just relies on the natural phenomena and their specific contents in plant tissue.

On the basis of the matrix of correlation coefficients, factor analysis was performed. The factor analysis was performed to identify and characterize element associations. From 16 analyzed variables (analyzed elements), 3 variables (Co, Ca and Mn), had very low factor loading, or tendency to form independent factor and therefore do not belong to any factor group. Nevertheless, these elements are generally present in the plant structure and are found in high percentages in soil, so they can be considered as geogenic elements. Three factors were identified, one anthropogenic and 2 geogenic, interpreted as Factor 1, Factor 2 and Factor 3, which cover 90% of variability of treated elements.

Factor 1 (Al-As-Cd-Cu-Fe-Pb-Zn): associates chemical elements that indicate anthropogenic influence in the study area (Figure 8). This association of elements was expected because of the geology of the study area and the works related to open ore pit and flotation activities. The acid drainage rapidly dissolved the elements, providing increased content in the soil. The open ore pit and flotation tailings dam allow direct exposure of the finest ore particles to the atmosphere. Corpuscle dust from the surface layer of the ore body and soil is spread in the atmosphere by the winds, in which way the atmospheric distribution of these elements in the vicinity of the mine is performed.

Factor 2 (Cr-Ni-Sr): presents typical geogenic factor (Figure 8). These elements are biogenic trace elements and are essential for the moss tissue. High factor loadings are related to the parts of Pleistocene sediments and Neogene dacites, andesites and pyroclastites. The correlation of Cr-Ni mostly relies on their geological occurrence in the study area. However, the Sr content correlation in this association also relies on certain natural phenomena.

Factor 3 (Ba-K-Na): these elements are naturally found in soil and moss as macro elements (Figure 10). The contents of these elements are variable and are not related to any anthropogenic activities. Their sources are mainly natural phenomena such as rock weathering and chemical processes in soil. This Factor is connected to the clay which is a product of disintegration mostly of feldspar and gneisses and micaschists. This means that the occurrence of this Factor is typical for the oldest formations in the Republic of Macedonia (Proterozoic micaschist and Proterozoic gneisses).

Table 1. Descriptive statistic of measurements for moss samples (values of Al, Ca, Fe, and K are given in % and the remaining elements in mg kg^{-1})

	n	Dis	X_a	X_g	Md	min	max	Var	s	CV
Al	52	log	2.12	1.83	1.72	0.47	8.51	1.77	1.33	62.8
As	52	log	2.62	1.54	1.55	0.14	13.7	9.36	3.06	117
Ba	52	N	32.4	29.7	30.6	11.5	66.0	166	12.9	39.7
Ca	52	log	6.43	6.30	6.24	4.53	10.6	1.87	1.37	21.3
Cd	52	log	0.54	0.48	0.49	0.18	1.75	0.08	0.28	51.2
Co	52	log	1.09	0.72	0.70	0.12	7.60	1.94	1.39	128
Cr	52	log	3.14	2.70	2.63	1.00	10.8	4.41	2.10	66.8
Cu	52	log	20.7	11.5	9.95	2.14	199	1141	33.8	163
Fe	52	log	3.29	2.85	2.63	0.74	12.4	3.86	1.96	59.7
K	52	log	3.22	3.17	3.16	1.93	4.51	0.30	0.54	16.9
Mn	52	log	165	153	144	59.0	439	4888	69.9	42.3
Na	52	log	46.3	44.6	45.8	25.1	81.5	164	12.8	27.6
Ni	52	log	7.37	6.53	6.21	2.10	30.1	21.0	4.58	62.1
Pb	52	log	8.82	7.36	6.81	2.68	40.2	46.1	6.79	76.9
Sr	52	log	26.1	24.7	24.1	12.9	55.4	83.0	9.11	34.9
Zn	52	log	29.2	28.4	28.3	17.3	53.7	46.5	6.82	23.4

Dis-distribution (log-lognormal; N-normal); X_a-arithmetic mean; X_g-geometric mean; Md-median; min-minimum; max-maximum; Var-variance; s-standard deviation; CV-coefficient of variance.

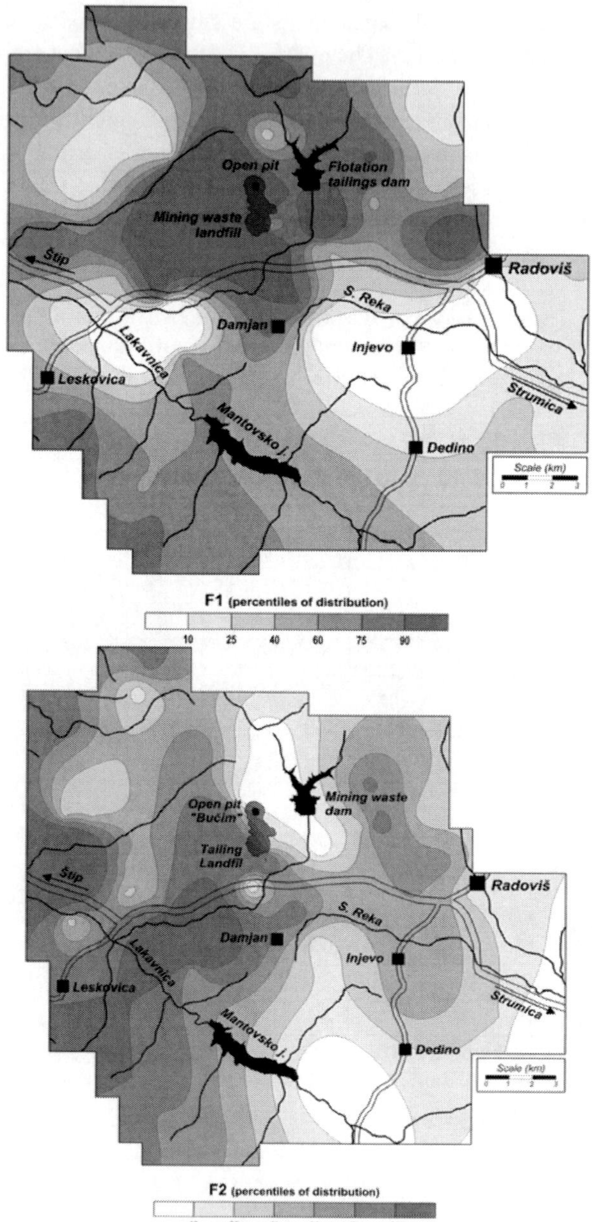

Figure 10. (Continued).

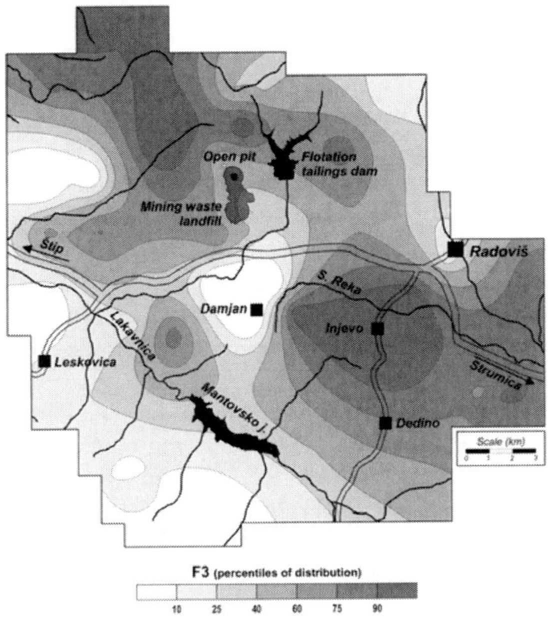

Figure 10. Areal distribution of F1, F2, F3 elements associations in moss species in Bučim copper mine environ (Balabanova et al., 2010).

4.2. Sasa Lead-Zinc Mine and Flotation (Case 2)

The descriptive statistics of the analysed elements is presented in Table 2. On the basis of normality tests and compared with histograms of distribution of the content of all analysed elements in moss samples, normality was assumed for the natural values of Co, Cr, K and Li. For the rest of the elements, the normality was assumed on the basis of the logarithms of their contents. However, it was established that the greatest part of the rest of the elements does not follow the Log-normal distribution (such as: As, Cd, Cu, Fe, Hg, Li, Mn, Ni, Pb, V, and Zn). The ultimate effect was achieved by applying Box-Cox data transformation. The power transformation used the principle method for dampening the difference between extreme values. High contents of Pb and Zn were detected (average values 60 and 75 mg kg^{-1}, respectively) in an area very close to the pollution source. Nevertheless, distribution of dust with high Pb-Zn contents was expected in the investigated area. Fine dust from flotation tailings is distributed, by the winds and deposited on moss tissue surface where accumulation of metallic ions occurs. Almost 0.05% of these heavy metals are accumulated in the moss plant tissue, confirming how good hyper-accumulators of heavy metals they are. Significant skewed distributions were the main problem in the data processing. This research has originally started with this approach, based on the assumption that the sampling medium base should not interfere with the natural distribution of element in the environment.

The investigated area represents an area where higher contents of certain heavy metals are expected, because it needs a medium (to monitor the content of the heavy metals contained in the finest dust which floats in the air), which actually represents a very good bio-indicator (Barandovski et al., 2008, 2012; Balabanova et al. 2010, 2013). Copper, with an average content of 11 mg kg^{-1}, was assumed to be anthropogenically introduced element in airborne particles, taking into account its value in comparison with the whole territory of the Republic of Macedonia (3.5 mg kg^{-1}) (Barandovski et al., 2013). The maximum value (56 mg kg^{-1}) indicates an anthropogenic impact to the Cu content in the moss tissue. The contents of Cr and Ni, as potentially toxic elements in higher concentrations, were established as only geologically introduced elements, due to the soil surface dusting (max. values 5.1 and 6.41 mg kg^{-1}, respectively). Mercury contents ranged from 0.021-0.08 mg kg^{-1}, showing no deviation from the Macedonian range (0.01-0.59) as it is presented by Barandovski et al. (2013). Opposite to this, Pb and Zn contents show anthropogenically introduced contents in the investigated area. The maximum values of 450 mg kg^{-1}, for both elements are perceived as enrichment of ~10 times for Pb and ~2 for Zn, compared with the whole territory of the country (Barandovski et al., 2013).

Four factor groups were revealed with the application of factor analysis: F1 (Al-Co-Cr-Fe-Li-Ni-V), F2 (Cd-Pb-Zn), F3 (Ca-Mg-Na-P) and F4 (Cu). The following elements: As, Ba, Hg, K, Mn, and Sr, were excluded from the further analysis because after the Box-cox data transformation, significant correlations were not assumed.

Factor 1 (Al-Co-Cr-Fe-Li-Ni-V): associates chemical elements that indicate geogenic distribution in the study area considering the generalized geology of the region. Contents of these elements are variable and are not related to any anthropogenic activities. Their sources are mainly natural phenomena such as rock weathering and chemical processes in soil. This factor is connected to the clay which is a product of disintegration mostly of feldspar and gneisses and micaschists. This means that the occurrence of this elements association is typical for the oldest formations in the Republic of Macedonia (Proterosoic shales and Proterosoic gneisses).

Factor 2 (Pb-Zn-Cd): the corpuscle dust from the surface layer of flotation tailings and soil is spread in the atmosphere by the winds, and in this way the atmospheric distribution of these elements in the vicinity of the mine is actually carried out. The areal distribution of these anthropogenic elements gives better visualization of the anthropogenic introduction of the relatively high contents of these toxic metals (max. values: 3.7 mg kg^{-1} for Cd and ~450 mg kg^{-1} for Pb and Zn). This element association presents an anthropogenic marker in the Sasa mine environment due to the Pb-Zn minerals deposition (Figure 11).

Table 2. Descriptive statistics for elements content values in moss samples (n=36, values given in mg kg^{-1})

	Xa	X_G	Md	min	max	s	CV
Al	3200	2500	2500	680	13000	2500	78
As	2.9	2.1	2.0	0.56	13	2.9	99
Ba	49	40	45	11	140	32	66
Ca	6600	6200	6500	2900	14000	2300	34
Cd	0.63	0.37	0.31	0.060	3.7	0.82	130
Co	0.72	0.56	0.53	0.16	2.6	0.58	81
Cr	2.3	2.0	2.1	0.84	5.1	1.1	47
Cu	11	8.3	7.2	3.6	57	12	104
Fe	3600	2800	2500	820	18000	3200	88
Hg	0.037	0.035	0.033	0.021	0.080	0.013	35
K	5200	4900	4600	2000	9800	1800	36
Li	1.2	0.98	1.1	0.31	3.9	0.70	61
Mg	3200	3100	3200	1700	4800	720	23
Mn	190	150	160	43	550	130	67
Na	74	46	41	20	890	150	199
Ni	2.9	2.7	2.7	1.1	6.4	1.2	40
P	88	830	770	410	1500	300	35
Pb	60	20	24	0.14	450	100	170
Sr	19	17	17	7.2	36	8.1	44
V	3.5	2.8	3.1	0.76	9.7	2.2	63
Zn	76	44	36	11	460	110	139

Dis. – distribution (Log – lognormal; N – normal); X – mean; X_G – geometric mean ; Md – median; min – minimum; max – maximum; s –standard deviation; CV – coefficient of variation.

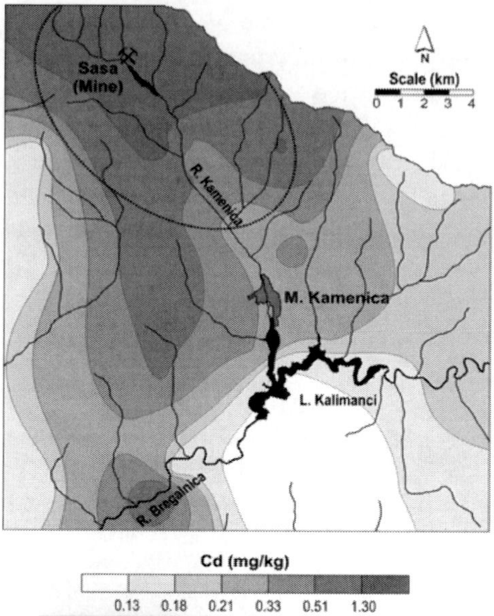

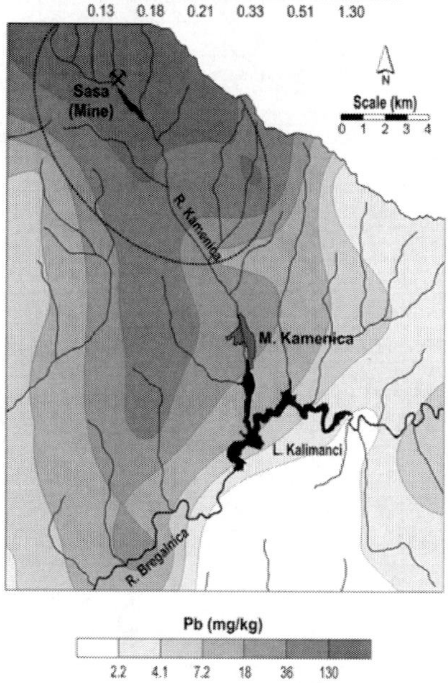

Figure 11. (Continued).

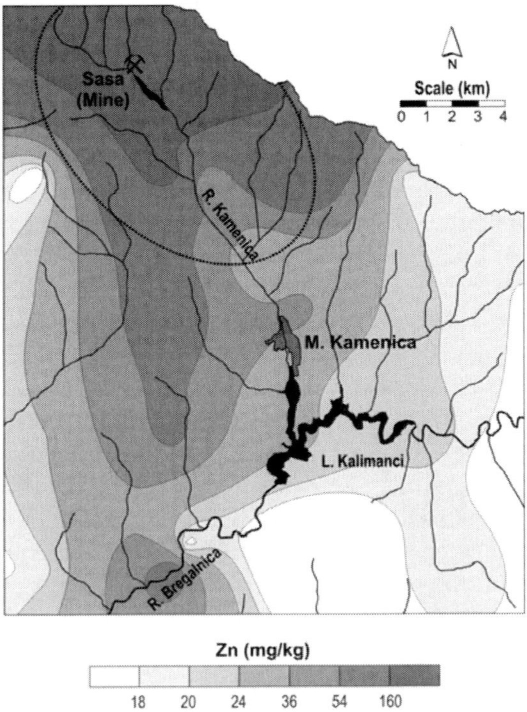

Figure 11. Areal distribution of Factor 1 (Pb-Zn-Cd) in the Sasa lead-zinc mine environ (Balabanova et al., 2013).

Factor 3 (Ca-Mg-Na-P): the occurrence of this factor is very similar to factor 1, resulting from natural phenomena. After Box-Cox data transformation, the influence of extreme values was reduced. Relatively higher contents are occurring due to the macroelements contents in the bryophyte tissue and intracellularly, depending on the elements bio-accumulation.

Factor 4 (Cu): presents a very interesting and specific factor because of the specifics of the copper distribution in the Pb-Zn ore deposited in the environment (neglecting the Eigenvalue, 0.97). There is a double impact on the Cu content in mosses. On one hand, it can be considered as a microelement essential for the plant tissue, but on the other hand, a relatively significant copper content can be detected, introduced from emissions from anthropogenic source, related to the dust distribution from the mine flotation tailings.

4.3. "Zletovo" Lead-Zinc Mine and Probištip Flotation Plant (Case 3)

All values of the metals' contents were statistically processed using descriptive statistics as presented in Table 3. Based on histogram plot for the following elements: Al, Cr, Fe, Mn, Ni, and V normal distributions were assumed. For the rest of the elements, normality was assumed on the basis of the logarithms of their contents. Higher contents of Pb and Zn were obtained than the average values of 34 and 46 mg kg^{-1}, respectively. The average values, in cases when anthropogenic enrichments are introduced in the environment, are not the best choice for comparative analysis. Thus, it would be preferable to use the median values as less influenced statistical parameters. The median values for the main anthropogenic elements were 15.4 and 33.8 mg kg^{-1}, respectively. Compared with the corresponding values referring to the whole territory of the R. Macedonia the values of enrichments were established based on the data obtained from Barandovski et al. (2013). For the Pb contents the large scale/small scale enrichments were calculated to be 3.35. For the Zn content the enrichment value was 1.7 (20 mg kg^{-1} for the whole territory of the R. Macedonia). In addition, the contents of Cd, Cu and Mn were also assumed as anthropogenically influenced from the emission source, based on their range values: 0.08-1.73 mg kg^{-1}; 4.11-21.4 mg kg^{-1}; and 55.5-376 mg kg^{-1}, respectively. The manganese distribution was assumed as normal based on the histogram visualization of the data set and comparison with the geologic maps of distribution. However, mine and flotation works that have influence as anthropogenic enrichment, provide another point of view. Median values were compared with the corresponding values for the whole territory of the R. Macedonia and there were no significant differences detected for the Cd and Mn. Only for the Cu contents, enrichments were found, and calculated to be ~2 times (Barandovski et al., 2013).

The values of the contents of plant-biogenic elements: Ca, K, Na, P, Mg, fall into the typical range for macro and micro elements in moss tissue. Their statistical data distributions were assumed as lognormal, as a consequence of the natural phenomena. Calcium content varies in the range of 0.3%-1.8%, for K: 0.2%-0.7%; for P: 0.05%-0.3% and for Mg: 0.1%-0.37%. Less enriched contents occurred for Na (23-78 mg kg^{-1}) and for Sr (17-123 mg kg^{-1}). There is no significant trend for variations in moss tissue biogenic elements compared with moss study issue for the whole territory of the R. Macedonia (Barandovski et al., 2013).

Table 3. Matrix of basic descriptive statistics parameters of the elements contents in moss samples (values of elements contents are given in mg kg^{-1})

	Dis	X_a	X_g	Md	min	max	s	CV
Ag	log	0.12	0.09	0.09	0.03	0.4	0.1	85.2
Al	N	3925	3578	3969	1530	7169	1686	42.9
As	log	2.35	1.67	1.67	0.5	8.35	2.11	89.7
Ba	log	68.4	61.2	58.8	21.3	226	39.8	58.2
Ca	log	10871	10338	10585	3708	17838	3272	30.1
Cd	log	0.31	0.23	0.21	0.08	1.73	0.35	114
Cr	N	3.47	2.99	3.44	0.51	7.43	1.72	49.8
Cu	log	9.28	8.34	6.82	4.11	21.4	4.88	52.5
Fe	N	3795	3405	3624	1345	8269	1777	46.8
K	log	3850	3666	3490	2231	7178	1336	34.7
Li	log	2.01	1.78	1.99	0.604	3.62	0.91	45.6
Mg	log	2127	1938	1602	1130	4367	1010	47.5
Mn	N	180	167	169	55.5	376	72.1	39.9
Mo	log	0.23	0.17	0.18	0.07	0.6	0.18	81.1
Na	log	45.7	42.5	44.4	22.8	78.2	17.6	38.5
Ni	N	4.07	3.63	3.75	0.94	11.1	2.08	51.1
P	log	1499	1392	1511	597	2930	563	37.5
Pb	log	33.8	18.9	15.4	4.01	200	50.6	149
Sr	log	44.3	39.3	36.5	18.6	123	24.1	54.2
V	N	8.20	7.36	8.11	2.16	15.8	3.57	43.6
Zn	log	46.5	37.8	33.8	12.8	186	39.1	84.2

Dis. –Distribution (log–lognormal; *N*–normal); X_a–arithmetic mean; X_g–geometric mean ; Md–median; min–minimum; max–maximum; *s*–standard deviation; *CV* – coefficient of variation.

The contents of the lithogenic elements Al, Cr, Fe, Mo, Ni, depend on the geology of the region. Aluminium content in the moss species varies in the range 0.15%-0.7% with a median value of 0.4%, and characteristic normal data distribution. As it has been previously presented by Barandovski et al. (2013) the trend of Al distribution on the whole territory of R. Macedonia varies in the range 0.05%-0.87%, and strongly relies on the lithology of the area (dominant magmatic and volcanic rocks, enriched with quaternary sediments). However, the median value of 0.2% suggests a lithologic introduction of Al in the study area. Iron content in moss tissue was accumulated in the range of 0.13%-0.83%. The Fe distribution follows the Al distribution, probably because of the same lithologic impact. For the rest of the lithologicically distributed elements Cr, Mo and Ni, the median values were in accordance with their normal distribution for the whole territory of the R. Macedonia (3.44, 0.18 and 3.75 mg kg^{-1}, respectively). It is relatively difficult to distinguish and determine the type of deposition and distribution based only on descriptive statistics of the contents of the elements in the investigated area.

In order to determine the correlations between the elements and the relationships between the elements contents, a matrix of correlation coefficients was processed. As lithologic elements, significant correlations were established for the following: Cr-Ni; Fe-Al; Cu-Ag; Al-V; Li-Cr; Ag-Zn. As regards the matrix elements in the plant tissue, the relations of the following elements were considered as significant: P-Ca; Mg-Na; Ba-Sr; Mg-Zn. Potentially hazardous relations of elements that are related to pollution emissions are: Pb-Zn; Cu-Zn; Cu-Pb; Cu-Cd; Cd-Zn; Mn-Zn.

Principal components and classification analysis was performed on the data set (elements contents) as a data reduction method used for description of supplementary variables and cases. In principal components analysis, after the first factor has been extracted, another line that maximizes the remaining variability should be defined. In this manner, consecutive principle components are extracted as PC1 and PC2 (using screen plot as extraction method). Because each consecutive factor is defined to maximize the variability that is not captured by the preceding factor, consecutive factors are independent of each other. Put in another way, consecutive factors are uncorrelated or orthogonal to each other. The first principal component PC1 associates the elements: Cu>Ag>Mg>Cd~Zn>Na>Pb~Mo, sequenced according to the expressions of component patterns. The second principle component PC2 associates the elements: Li>V>Al>Cr>Ca~As. The elements Ba, Fe, K, Mn, Ni, P and Sr, do not belong to any group, because of the very low component pattern for the PC1 and PC2.

4.4. "Fe-Ni Industry" Ferro-Nickel Smelter Plant Case Study (Case 4)

The descriptive statistics of the 46 elements analyzed in the moss samples collected from the investigated region is given in Table 4. The analyses are based on the nitric acid solution's potential to leave out fractions of the elements in the moss samples contained in silicate minerals (attached to the soil particles). The median values and min-max ranges for the anthropogenic groups of elements in the moss samples from the investigated region (n = 31) are presented in Table 4, together with the corresponding data from similar studies on the territory of the Republic of Macedonia (Barandovski et al., 2013). In the further text, data about the toxic elements will be presented and discussed. For the rest of the elements there was no such significant difference between the median values for the moss samples from the investigated region and the moss samples collected from the entire territory of the Republic of Macedonia, as well as moss samples in Europe from a survey of 2010/2011 (Harmens et al., 2013). Most emissions of *nickel* to the environment originate from local sources, mainly from mines and smelters. Nickel is essential to some organisms, but it is also toxic in higher concentrations to most plants, fungi and animals. It can also be dangerous when the maximum tolerable amounts are exceeded. Nickel content ranges from 14.0 to 340 mg kg^{-1}, with a median value of 40 mg kg^{-1}. The median value is much higher than the nickel median value of 3.5 mg kg^{-1} (ranging from 1.3 to 52 mg kg^{-1}) in samples taken throughout the whole territory of the Republic of Macedonia (Barandovski et al., 2013). It is evident from these results that the values obtained for Ni content are also significantly higher than the European values from the 2010 moss survey where the median value of Ni is stated to be 1.94 mg kg^{-1} (Harmens et al., 2013). If we compare the median values for nickel in the moss samples collected from other countries in the European moss network from the Balkan region (Bulgaria, Croatia, Slovenia) (2.61, 3.16, and 2.12 mg kg^{-1}, respectively), as stated in a survey of 2010/2011, it is evident that the Republic of Macedonia has the highest value of 3.45 mg kg^{-1}.

This data shows that this element is significantly affected by the anthropogenic activities in the ferronickel smelter plant located near the town of Kavadarci. Coal and heavy oil combustion as well as chromium smelter plants are the main sources of increasing *chromium* deposition in the surrounding ecosystems. Chromium content in the mosses in the Kavadarci region varies between 5.8 to 110 mg kg^{-1}, with a median of 15 mg kg^{-1} which is about ~3.0 times higher than the median values for Cr content in the moss samples taken from the whole territory of Macedonia in 2010 (median of 3.5 mg kg^{-1}) and 4.73 times higher than the median for the European mosses collected in 2010/2011 (median of 1.82 mg kg^{-1}), as reported by Barandovski (2013) and Harmens (2013). The main source of cobalt in the environment is the steel industry, and it is also produced as a by-product in copper and nickel mining. The median value for Co contents in the moss samples taken from the studied area is 2.1 mg kg^{-1} (ranging from 1.1 to 14.0 mg kg^{-1}).

Table 4. Descriptive statistic of measurements for moss samples (n = 31, values in mg kg^{-1})

Element	Min	Max	Md	X_g	X	s	CV
Ag	0.030	0.70	0.12	0.11	0.16	0.16	102
Al	840	5400	1800	1900	2200	1200	56
As	0.10	6.2	0.92	0.90	1.9	2.1	112
Au	0.005	0.37	0.034	0.023	0.047	0.072	154
Ba	19	97	39	41	47	25	53
Be	0.025	2.3	0.18	0.14	0.31	0.47	153
Ca	6300	15000	11000	10000	11000	2400	23
Cd	0.040	0.51	0.19	0.18	0.21	0.11	54
Ce	1.89	12	4.7	5.2	5.8	3.0	51
Co	1.08	14	2.1	2.8	3.7	3.2	84
Cr	5.76	112	15	19	28	27	97
Cs	0.15	1.7	0.34	0.39	0.46	0.33	70
Cu	4.94	15	7.7	8.0	8.3	2.4	29
Dy	0.10	0.87	0.29	0.30	0.35	0.20	58
Er	0.020	0.47	0.15	0.14	0.17	0.098	57
Eu	0.050	0.26	0.10	0.11	0.12	0.056	47
Fe	1300	8700	2300	2700	3200	1900	60
Ga	0.36	2.9	0.93	1.0	1.1	0.62	54
Gd	0.16	1.3	0.43	0.45	0.52	0.30	57
Ge	0.005	0.12	0.060	0.049	0.063	0.035	56
Hg	0.005	0.090	0.060	0.051	0.057	0.021	36
Ho	0.010	0.16	0.050	0.054	0.064	0.038	59
K	2700	10000	5500	5500	5800	1800	31
La	0.77	5.4	1.8	2.0	2.4	1.3	57
Li	0.98	8.2	2.0	2.3	2.6	1.6	61
Lu	0.005	0.050	0.015	0.016	0.019	0.012	67
Mg	1300	7000	3300	3400	3600	1300	35
Mn	54	300	140	140	150	60	41
Mo	0.025	10	0.79	0.48	1.6	2.4	152
Na	50	1000	310	300	370	230	62
Nb	0.070	1.4	0.25	0.24	0.30	0.25	82
Ni	14	340	40	48	80	89	112
P	990	5600	3000	2800	3000	990	34
Pb	5.40	19	8.4	9.1	9.7	3.6	37
Rb	4.16	31	7.7	8.4	9.4	5.4	57
Sb	0.030	1.6	0.090	0.20	0.44	0.48	110
Sm	0.11	1.1	0.41	0.42	0.48	0.26	54
Sr	11	130	39	36	44	29	68

Element	Min	Max	Md	X_g	X	s	CV
Tb	0.020	0.18	0.050	0.053	0.064	0.042	66
Th	0.23	2.4	0.62	0.71	0.84	0.52	62
Ti	46	400	150	140	170	95	57
U	0.11	0.81	0.16	0.20	0.24	0.18	74
V	2.65	23	7.6	8.0	9.1	5.1	55
Yb	0.030	0.32	0.10	0.10	0.12	0.075	62
Zn	30	190	55	64	74	46	62
Zr	0.025	4.0	0.93	0.75	1.3	1.0	83

Min – minimum; Max – maximum; Md – median; X_g – geometric mean; X –mean; s – standard deviation; s_x – standard error of mean; CV – coefficient of variation (%).

Lead, at certain exposure levels, is a poisonous substance to animals. Considering our results, it was found that the presence of lead in the environment in the Republic of Macedonia is mainly connected with the lead-zinc mines and smelter operations. In this study it was found that the lead content in the mosses in the region of the ferronickel smelter plant near Kavadarci ranges from 5.4 to 19.0 mg kg^{-1}, with a median of 8.4 mg kg^{-1}. This is 2 times higher than the median value for lead (4.6 mg kg^{-1}) in the moss samples from all over Macedonia collected in 2010 (Barandovski et al., 2013).

The influence of Pb-Zn smelter plants situated in the city of Veles (Stafilov et al., 2010) is very obvious if we compare the obtained values, those for Macedonia and the European median values with those for Bulgaria, where the median value is 8.0 mg kg^{-1} as reported in the survey from 2010/2011 (Barandovski et al., 2013).

Zinc is an essential micronutrient in all organisms, but at higher concentrations it could be moderately toxic to plants, and only slightly toxic to mammals. Zinc concentration in the mosses in the vicinity of Kavadarci ranges between 30 and 190 mg kg^{-1} with a median of 55 mg kg^{-1}, which is about 2.7 times higher than the median value for zinc for the whole territory of the Republic of Macedonia (20 mg kg^{-1}) in the surveys conducted in 2010.

On the basis of the matrix of correlation coefficients, factor analysis was carried out. Principal component factor analysis was used to identify and characterize element associations [23] Three factors were identified, one geogenic (Al-Ce-Cs-Dy-Er-Eu-Ga-Gd- Ho-La-Li-Lu-Sm-Tb-Th-Ti-V-Yb) and two anthropogenic (Ni-Co-Cr-Cu-Fe-Mg and As-Cd-Cu-Hg-Pb-Zn) associations, interpreted as Factor 1, Factor 2 and Factor 3. The existing literature data confirms that the elements belonging to the first anthropogenic association have relatively higher contents in the iron-nickel ore used in the production of nickel in the smelter plants (between 1 and 2.5% Ni, about

0.05% Co and 1–3% Cr), present mainly in silicate forms. The distribution of Ni, Cr and Co presents a specific anthropogenic marker for the Fe-Ni smelter plant operation and local air pollution. The areal distribution of these elements in the Kavadarci environment is presented in Figure 12.

4.5. Allchar As-Sb-Tl Mine (Case 5)

The data obtained by ICP-MS and ICP-AES was based on nitric acid solutions, possibly leaving out fractions of the elements contained in silicate minerals in the precipitate (soil particles). In Table 5, the values of Al, Ca, K, Mg and P are given in %, and the values of Ag, Be, Bi, Dy, Er, Eu, Gd, Ge, Hf, Hg, Ho, Sb, Sm, Tb, W and Yb are given in μg kg^{-1} and for the remaining elements in mg kg^{-1}. On the basis of the normality tests and the comparison with histograms of distribution for the contents of all analysed elements, normality was assumed for the natural values of Ge, I, Mg and P. For the rest of the elements, the distribution was assumed on the basis of the logarithms of their contents. The median values obtained in this study were compared with those obtained in the moss studies in the Republic of Macedonia carried out in 2002, 2005 and 2010 (Barandovski et al., 2008, 2012, 2013). The median values for As in the samples collected very close to the mine (2.1 mg kg^{-1}) were compared to those referring to the whole territory of the R. Macedonia which are as follows: for 2002 - 0.8 mg kg^{-1}, for 2005 - 0.67 mg kg^{-1} and for 2010 - 0.50 mg kg^{-1}. It can be concluded that there is certain enrichment of the As content in the moss samples from the area very close to the mine. However, the arsenic value for the whole investigated area (0.49 mg kg^{-1}) is not a reason for concern. The antimony content (0.029 mg kg^{-1}) does not show deviation in enrichment from the respective values for the whole territory of the R. Macedonia from the three successive investigations in 2002, 2005 and 2010 (0.20, 0.15 and 0.090 mg kg^{-1}, respectively). On the other hand, the thallium content in the moss samples collected very close to the mine (0.96 mg kg^{-1}) is enriched in comparison with the respective values for the moss species from the whole territory of the R. Macedonia (<0.10 mg kg^{-1} in all three years of investigation).

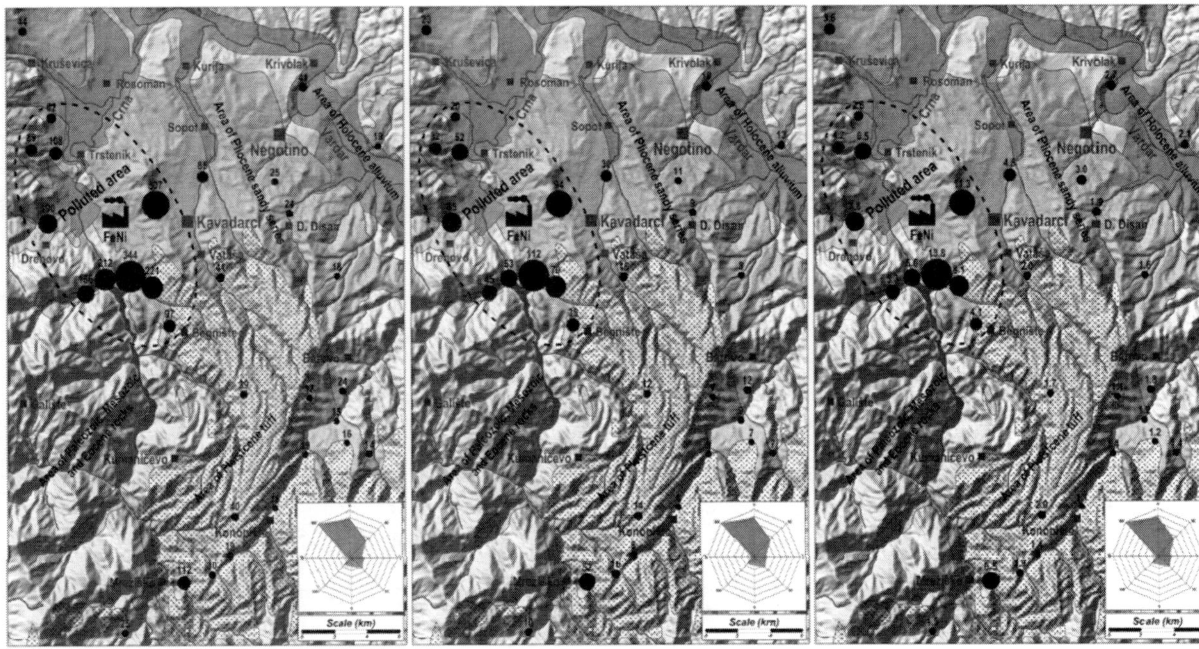

(a) (b) (c)

Figure 12. Areal distribution of Ni (a), Cr (b) and Co (b) in “FENI” ferro-nickel smelter plant environ (Bačeva et al., 2013).

Table 5. Descriptive statistics for the values of 53 elements content in the moss samples (n=69). Data rounded to two digits

	Unit	Dis	X	Xt	Xg	Md	Min	Max	P_{25}	P_{75}	S	S_X	CV	A	E
Ag	μg kg^{-1}	Log	25	22	22	20	10	140	16	28	18	2.1	72	1.24	2.73
Al	%	Log	0.22	0.17	0.15	0.13	0.029	1.4	0.097	0.20	0.28	0.033	127	1.03	1.45
As	mg kg^{-1}	Log	2.8	2.0	0.68	0.49	<0.1	30	0.18	2.4	5.5	0.67	197	0.34	-0.67
Ba	mg kg^{-1}	Log	38	33	24	19	3.6	200	12	53	41	5.0	110	0.26	-0.42
Be	μg kg^{-1}	Log	57	45	35	30	<10	370	20	49	78	9.4	138	0.61	1.08
Bi	μg kg^{-1}	Log	20	18	16	15	<10	88	11	22	15	1.8	76	0.08	0.12
Br	mg kg^{-1}	Log	0.55	0.50	0.29	0.43	<0.01	3.4	0.20	0.69	0.53	0.064	96	-1.61	1.99
Ca	%	Log	0.58	0.58	0.57	0.57	0.29	0.86	0.50	0.68	0.12	0.015	21	-0.55	0.06
Cd	mg kg^{-1}	Log	0.15	0.15	0.13	0.13	0.038	0.67	0.095	0.19	0.094	0.011	61	0.40	0.44
Ce	mg kg^{-1}	Log	1.4	1.2	0.99	0.89	0.17	9.6	0.61	1.4	1.7	0.20	118	0.63	0.83
Co	mg kg^{-1}	Log	0.87	0.79	0.76	0.69	0.29	3.4	0.57	0.91	0.59	0.071	67	1.09	1.84
Cr	mg kg^{-1}	Log	5.7	5.3	5.1	5.0	2.2	20	3.8	6.3	3.6	0.43	62	1.08	1.83
Cs	mg kg^{-1}	Log	0.60	0.37	0.18	0.14	0.021	7.7	0.075	0.33	1.4	0.16	225	1.01	0.55
Cu	mg kg^{-1}	Log	3.2	2.9	2.8	2.7	0.021	11	2.1	3.3	1.9	0.22	58	1.27	1.66
Dy	μg kg^{-1}	Log	110	93	80	72	17	870	51	110	130	16	117	0.85	1.03
Er	μg kg^{-1}	Log	57	47	40	36	<10	420	25	57	66	8.0	116	0.64	1.21
Eu	μg kg^{-1}	Log	35	29	24	22	<10	280	15	36	41	4.9	116	0.32	0.61
Fe	mg kg^{-1}	Log	1200	930	870	730	230	6800	600	1200	1300	150	110	1.24	2.30
Ga	mg kg^{-1}	Log	0.44	0.33	0.30	0.25	0.068	3.5	0.19	0.42	0.61	0.073	138	1.27	2.25
Gd	μg kg^{-1}	Log	150	130	110	98	23	1200	69	160	180	22	117	0.76	0.82

	Unit	Dis	X	Xt	Xg	Md	Min	Max	P_{25}	P_{75}	S	S_X	CV	A	E
Ge	μg kg^{-1}	N	15	14	13	14	<10	33	11	18	7.4	0.90	51	0.65	0.01
Hf	μg kg^{-1}	Log	23	20	17	17	<10	140	12	24	21	2.5	91	0.15	0.35
Hg	μg kg^{-1}	Log	74	65	48	40	<10	410	26	82	81	9.7	109	0.29	0.12
Ho	μg kg^{-1}	Log	21	17	14	14	<10	160	<10	22	25	3.0	123	0.62	0.04
I	mg kg^{-1}	N	0.21	0.20	0.19	0.20	0.033	0.38	0.16	0.25	0.073	0.009	36	0.08	-0.24
K	%	Log	0.46	0.46	0.44	0.45	0.22	0.83	0.37	0.54	0.14	0.016	30	0.14	-0.10
La	mg kg^{-1}	Log	0.69	0.52	0.45	0.40	0.081	6.3	0.26	0.61	1.0	0.12	149	0.91	1.81
Li	mg kg^{-1}	Log	1.0	0.74	0.70	0.60	0.15	13	0.49	0.85	1.7	0.20	165	1.58	4.87
Mg	%	N	0.11	0.11	0.11	0.11	0.077	0.16	0.093	0.12	0.022	0.003	19	0.43	-0.66
Mn	mg kg^{-1}	Log	71	65	57	62	16	330	34	88	54	6.5	76	0.18	-0.23
Mo	mg kg^{-1}	Log	0.20	0.18	0.17	0.15	0.058	1.1	0.12	0.20	0.18	0.022	88	1.11	1.44
Na	mg kg^{-1}	Log	140	120	100	91	44	850	77	110	160	19	113	2.07	4.21
Nd	mg kg^{-1}	Log	0.71	0.59	0.49	0.45	0.083	5.0	0.31	0.70	0.83	0.10	117	0.64	0.79
Ni	mg kg^{-1}	Log	13	12	12	11	6.5	27	9.7	14	4.5	0.54	35	0.56	0.29
P	%	N	0.10	0.10	0.096	0.11	0.023	0.18	0.077	0.13	0.033	0.004	32	-0.01	0.01
Pb	mg kg^{-1}	Log	11	4.4	3.9	4.0	<0.01	480	3.2	5.8	57	6.9	501	-2.69	26.46
Pr	mg kg^{-1}	Log	0.17	0.14	0.12	0.11	0.021	1.2	0.076	0.17	0.20	0.024	117	0.63	0.82
Rb	mg kg^{-1}	Log	5.6	5.0	4.4	4.3	1.6	26	2.6	6.3	4.7	0.57	85	0.72	0.10
S	mg kg^{-1}	Log	680	660	640	580	290	1600	520	700	270	33	40	1.19	4.28
Sb	μg kg^{-1}	Log	25	16	8.2	10	<10	330	<10	28	51	6.1	202	0.70	-0.83
Sc	mg kg^{-1}	Log	0.49	0.42	0.40	0.35	0.17	2.6	0.30	0.47	0.44	0.053	91	1.59	3.11
Sm	μg kg^{-1}	Log	130	110	86	<100	<100	1100	<100	150	170	21	130	1.29	0.62
Sn	mg kg^{-1}	Log	1.7	0.86	0.89	0.74	0.42	47	0.65	0.94	5.6	0.67	338	3.78	18.52

Table 5. (Continued)

	Unit	Dis	X	Xt	Xg	Md	Min	Max	P_{25}	P_{75}	S	S_X	CV	A	E
Sr	mg kg^{-1}	Log	21	18	15	12	4.7	120	8.5	19	23	2.7	108	0.97	0.19
Tb	µg kg^{-1}	Log	20	16	13	13	<10	160	<10	22	25	3.0	124	0.66	-0.07
Th	mg kg^{-1}	Log	0.45	0.37	0.28	0.26	<0.1	3.2	0.17	0.48	0.57	0.069	126	0.34	0.48
Tl	mg kg^{-1}	Log	4.8	0.44	0.20	0.12	<0.1	290	<0.1	0.77	35	4.2	740	1.51	3.56
V	mg kg^{-1}	Log	2.8	2.3	2.1	1.8	0.51	17	1.4	2.8	3.1	0.37	110	1.07	1.67
W	µg kg^{-1}	Log	21	19	16	15	<10	95	11	28	16	2.0	78	0.05	-0.34
Y	mg kg^{-1}	Log	0.55	0.46	0.40	0.35	0.091	4.0	0.25	0.54	0.63	0.076	114	1.01	1.29
Yb	µg kg^{-1}	Log	49	40	34	31	<10	360	21	49	57	6.9	117	0.56	1.19
Zn	mg kg^{-1}	Log	17	17	17	16	10	32	14	19	4.3	0.52	25	0.21	-0.20
Zr	mg kg^{-1}	Log	0.76	0.69	0.64	0.60	0.19	3.3	0.44	0.78	0.54	0.065	71	0.69	0.81

Dis. – distribution (*N* – normal, *Log* – lognormal;); *X* – mean; X_t – trimmed mean (5%); X_g – geometric mean; Md–median; Min – minimum; Max–maximum; P_{25}– lower quartile; P_{75} – upper quartile; S – standard deviation; S_x – standard error of mean ;*CV* – coefficient of variation; *A* – skewness; *E* – kurtosis.

On the basis of the matrix of correlation coefficients, cluster and factor multivariate statistical analysis was performed. Factor analysis was used to identify and characterize element associations. From 53 analysed variables (analysed elements), 12 analysed variables Br, Ca, Hg, I, K, Na, P, Pb, S, Sn, Th and W were eliminated from further analysis, because of their tendency to form their own clusters, not showing a reasonable connection to other chemical elements. Elements with low share of communality or tendency to form independent factors were also excluded. The matrix of rotated factor loadings is presented and five factors are identified in Table 4: **F1** (Co, Cr, Fe, Sc, Li, V, Ga, Y, Ni, Mn, Al, La-Lu, Cu, Ge, Be, Bi, Hf); **F2** (As, Tl, Sb, Mg); **F3** (Rb, Cs, Mo); **F4** (Sr, Ba, Hf, Zr, La-Lu, Bi) and **F5** (Cd, Zn, Ag, Cu) associations, interpreted as Factors (F1 - F5), which account for 81 % of the total variability of treated elements.

Factor 1 (F1) is the strongest factor, representing 36.2% of the total variability. F1 associates the following elements: Co, Cr, Fe, Sc, Li, V, Ga, Y, Ni, Mn, Al, La-Lu, Cu, Ge, Be, Bi and Hf. The group includes chemical elements that are probably naturally distributed. The geographical distribution of F1-association is shown in Fig. 5A. High concentrations of most elements of the F1 association are typical for Plio-Quaternary tuff and latite breccias and extremely low values for Pliocene andesitic. The scale refers to the wider area, and therefore it cannot be interpreted locally. The regional distribution of the majority of these elements is typical for the group of crucial elements predominantly supplied to the moss by windblown soil dust. The relative enrichment in La is characteristic for the volcanic rocks of Allchar, similar to the Ce content, as well as Ce/Y (Boev and Jelenković 2012). The global trend of the value of F1 is the SE–NW, which is the influence of weak local topology.

Factor 2 (As, Tl, Sb and Mg): represents element association which indicates natural and anthropogenic influence in the study area. The distribution map for this geochemical association is given in Figure 13. From the distribution map of the F2 association, it can be clearly seen that higher contents of these elements are deposited in the close vicinity of the mine. This association of elements was expected because of the mining activities in the past (ore and tailings waste are present in the open) and as a result of surface phenomena in this area in the past (Stafilov et al., 1988; Janković, 1993; Frantz et al., 1994). In the case of the F2 distribution elements were very well defined, although this distribution is not directly related to geology. For this association the influence of local topology is also low. The local thermal winds carry up and down the valley the primary fine particles of dust from the surface layer and soil with high concentrations of As, Tl, Sb, and dispose them to the moss samples. This geochemical association of As-Sb-Tl is characterized by relatively high correlation coefficients between these elements. The association between As and Sb is weaker but also highly significant. Their source is mainly from natural phenomena. In the Allchar locality it might be better to replace the term “pollution” with the term "natural enrichment" which occurred by the erosion of the ore bodies. The real pollution that comes from the mine or mine tailings from the past was insignificant, which means that the presence of these elements in the Allchar area is mainly natural.

Factor 3 (Rb, Cs and Mo): high concentrations of these elements are present in Plio-Quaternary tuff outcrops, Triassic carbonates, there are relatively high concentrations in Pliocene andesites and very low concentrations in Quaternary moraines. The cause of this distribution cannot be established with certainty. Perhaps in question is remote impact, as in the case of the F1 association.

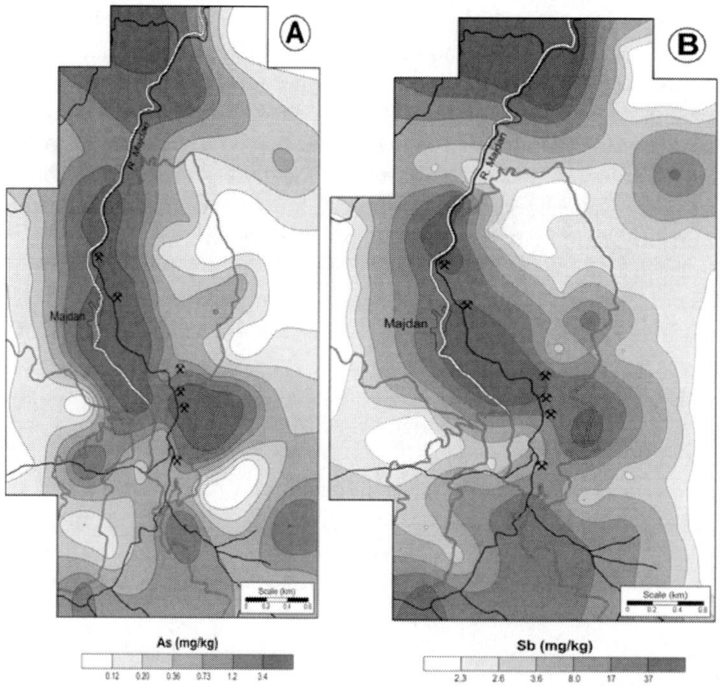

Figure 13. (Continued).

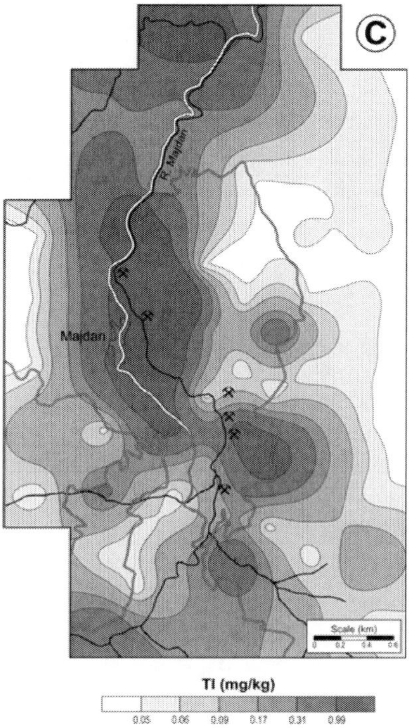

Figure 13. Areal distribution of F2 scores, As (a), Sb (b) and Tl (c) in Allchar As-Sb-Tl mine environ (Bačeva et al., 2013).

Factor 4 (Sr, Ba, Hf, Zr, La-Lu, Bi): this association is actually locally distributed. High values were related to quartz outcrops PLQ Latite Breccia (N and E of the study area) and a small part of the outcrops is related to Pl andesites. Low values of this group were found in the area of Jurassic diabase and serpentines (SW part). The contents of these elements that are reflection of the natural processes represented very rarely or not at all in industrial processes, depend mostly on the basic geological structure. Such is the example of the Ba content, which occurs as a natural marker in piroclastites and its tuffs.

Factor 5 (Cd, Zn, Ag, Cu) represents an association of elements that correlate typical "heavy metals", assumed to be the reason for the local pollution. It can be inferred that there is certain enrichment in regard to the heavy elements. This enrichment of heavy metals leads to increased content of these heavy metals in the atmosphere, primarily, because of the direct exposure of ore particles to the atmosphere. The presence of high contents of these heavy metals in the atmosphere results from the relative enrichment of Cu and Zn in some ore from the Allchar deposit (Frantz et al. 1994; Boev and Jelenković 2012). The fine particles from the ore are spread in the atmosphere, carried by the wind. High values of quartz outcrops were found in Plio-Quaternary latite breccias, Jurassic diabase and serpentinite. Low values were at Triassic clastites (N and E part).

Conclusion

The main goal of this study is to present an overview of small-scale monitoring at four microlocations in the Republic of Macedonia. Mosses are a very useful tool for determining the deposition of metals in the environment. Their relative usefulness and necessity in this type of analysis is proven by many studies. Therefore, on one hand moss can be used as a medium in wide areas to determine the hotspots of pollution and on the other hand, it tightens and thoroughly determines the extent of contamination and distribution at local level (very close environment of emission pollutants).

Precisely defined sampling networks were constructed at four microlocations on the territory of the R. Macedonia. Large scale monitoring was conducted on the whole territory of Macedonia, which helped localize these microlocations as emission pollutants. Therefore, the mosses were used as most suitable sampling media for determining the deposition and distribution of Ag, Al, As, Ba, Be, Bi, Ca, Cd, Ce, Co, Cr, Cs, Cu, Dy, Eu, Fe, Ga, Gd, Ge, Hf, Hg, Ho, I, K, La, Li, Lu, Mg, Mn, Mo, Na, Nb, Ni, Pb, Rb, Sb, Sm, Sr, Tb, Th, Ti, U, V, Yb, Zn and Zr. The following moss species: *Hypnum cupressiforme* (Hedw.) and *Camptothecium lutescens* (Hedw.), *Homolothecium sericeum* (Hedw.), and *Scleropodium purum* (Hedw.) Limpr. were established as characteristic plant species in the flora of Macedonia. The most frequent species was *Hypnum cupressiforme.*

The monitoring at the first microlocation in the Bučim copper mine and flotation plant determined increased content of certain heavy metals in the atmosphere. The application of statistical multivariate analysis (factor analysis) singled out one anthropogenic group of elements consisted of As, Cd, Cu, Fe, Pb and Zn. The presence of high contents of these heavy metals in the atmosphere has an impact on population health. The presence of the copper mine leads to increased content of these heavy metals in the atmosphere, primarily because of the direct exposure of ore particles and flotation tailings to the atmosphere (max. value of Cu content of 200 mg kg^{-1}). The fine particles from the ore are spread in the atmosphere carried by the wind. Microparticles penetrate into the human body by inhalation through the respiratory system. The maps of areal deposition of heavy metals in the area around the Bučim mine provide evidence that these metals are found in increased contents in the close vicinity of the mine. Distribution of these heavy metals in the distant parts of the study area was very low. However, there are two settlements in the close vicinity of the mine (villages Bučim and Topolnica) where the population is directly exposed to the presence of these heavy metals in the atmosphere, which certainly is an issue that needs attention.

At the second microlocation in the Sasa lead-zinc mine, the distribution of 21 elements was reduced to four synthetic associations: Al-Co-Cr-Fe-Li-Ni-V; Cd-Pb-Zn; Ca-Mg-Na-P; and Cu. There is a double impact on the Cu content in mosses. The background content in the moss tissue significantly interferes when anthropogenic introduction occurs. The anthropogenic association of elements (Cd-Pb-Zn) shows positive functional interdependence with the distance from the emission source. The anthropogenic elements (Cd, Pb and Zn) occur in potentially hazardous contents (max. values 3.7, 450, 460 mg kg^{-1}, respectively). The areal distribution confirms the anthropogenic influence of the mine on the close environment.

The results obtained from the Zletovo lead-zinc mine environment reveal a potentially polluted area, due to the maximum values obtained for the Pb and Zn (123 and 186 mg kg^{-1}). The results also suggest that smaller scale local deposition patterns are variable and could be completely examined using moss sampling media. The smaller scale monitoring of the emissions from the lead-zinc mine and flotation tailings is carried out following the wind directions in the investigated area. Moss species variability in elements accumulation was not found.

The obtained results from the ferro-nickel smelter plant environment (near the town of Kavadarci) unequivocally confirm the biomonitoring ability of the mosses in air pollution monitoring. It was found that there is an evidently higher content of nickel in the moss samples in the vicinity of the ferronickel smelter plant near Kavadarci. The median value of Ni in the moss samples from the whole region (40 mg kg^{-1}) is much higher than the median for Macedonia (5.82 mg kg^{-1}). Moreover, the median value for the nickel content in the moss samples from the polluted area (around the smelter) is 178 mg kg^{-1} with enrichment ratio in the moss samples almost 5.5 times higher than the unpolluted areas (32 mg kg^{-1}). This fact confirms the influence of the dust from the ferronickel plant on the air pollution in the wider region of the town of Kavadarci. According to the results of the statistical tests of the moss samples, an anthropogenic group of elements was determined in this area (Cd, Co, Cr, Fe, Mg, Ni, Pb and Zn). The distribution of these elements (especially Ni, Co and Cr) shows an increased content in the moss samples from the smelter plant surroundings compared with their content in the rest of the samples. Namely, the activities implemented in relation to the ferronickel ore processing and the smelter plant result in increased content of certain heavy metals in the atmosphere, which was determined based on the monitoring of moss samples.

The obtained values for 53 analyzed elements from the samples collected around the Allchar mine in the Allchar area show that the highest median values for As, Sb and Tl are detected in the vicinity of the mine . The atmospheric deposition of As in the moss samples from around the Allchar mine is >6.5 times higher and the value for Tl is 19 times higher compared with the values for the samples from the rest of the Allchar area. By comparing the results of this study with those of 2002, 2005 and 2010, it can be concluded that the situation in relation to air deposition of the investigated elements in the Allchar area compared with the moss survey for the whole territory of the Republic of Macedonia shows no significant variation except for As and Tl contents. This might be the result of the mining activities in the past (the presence of ore and tailings) and of past surface phenomena in this area. By application of Multivariate cluster and R-mode factor analyses (FA), five geochemical associations were determined: **F1** (Co, Cr, Fe, Sc, Li, V, Ga, Y, Ni, Mn, Al, La-Lu, Cu, Ge, Be, Bi and Hf); **F2** (As, Tl, Sb and Mg); **F3** (Rb, Cs and Mo); **F4** (Sr, Ba, Hf, Zr, La-Lu and Bi) and **F5** (Cd, Zn, Ag and Cu). The essence of this study, as it was expected, is the successful detection of the strongly expressed association of As-Sb-Tl, and also of the less prominent associations based on moss biomonitoring and multivariate statistical methods for small area.

As it can be summarized, moss species are readily available and can be efficiently analysed for certain heavy metals. The collected moss species show landscape and canopy-dependent accumulations that generally correspond to the expected concentrations. The results also suggest that smaller scale local deposition patterns are variable and could be thoroughly studied using moss sampling media. The smaller scale monitoring of the emissions from the lead-zinc mine and flotation tailings is carried out following the wind directions in the investigated area. Moss species variability in elements accumulation was not found.

REFERENCES

Aboal, J. R., Fernández, J. A., Boquete, T. & Carballeira, A. (2010). Is it possible to estimate atmospheric deposition of heavy metals by analysis of terrestrial mosses? *Science of the Total Environment*, *408*(24), 6291 - 6297.

Bačeva, K., Stafilov, T. & Šajn, R. (2009). Biomonitoring of nickel air pollution near the city of Kavadarci, Republic of Macedonia. *Ecology and Protection of the Environment*, *12*(1-2), 57-69.

Bačeva, K., Stafilov, T., Šajn, R. & Tănăselia, C., (2012). Moss biomonitoring of air pollution with heavy metals in the vicinity of a ferronickel smelter plant. *Journal of Environmental Science and Health, Part A: Toxic/Hazardous Substances and Environmental Engineering*, *47*(4), 645-656.

Bačeva, K., Stafilov, T., Šajn, R., Tănăselia, C. & Ilić Popov, S. (2011). Distribution of chemical elements in attic dust in the vicinity of ferronickel smelter plant. *Fresenius Environmental Bulletin*, *20*(9), 2306-2314.

Bačeva, K., Stafilov, T., Šajn, R. & Tănăselia, C. (2013). Air dispersion of heavy metals in vicinity of the As-Sb-Tl abandoned mine and responsiveness of moss as a biomonitoring media in small scale investigations. *Environmental Science and Pollution Research*, *20*(12), 8763-8779.

Balabanova, B., Stafilov, T., Bačeva, K. & Šajn, R. (2010). Biomonitoring of atmospheric pollution with heavy metals in the copper mine vicinity located near Radoviš, Republic of Macedonia. *Journal of Environmental Science and Health, Part A, Toxic/Hazardous Substance & Environmental Engineering*. *45*, 1504–1518.

Balabanova, B., Stafilov, T., Šajn, R. & Bačeva, K. (2011). Distribution of chemical elements in attic dust as reflection of their geogenic and anthropogenic sources in the vicinity of the copper mine and flotation plant. *Archives of Environmental Contamination and Toxicology*, *61*(2), 173-184.

Balabanova, B., Stafilov, T., Šajn, R. & Bačeva, K. (2012) Characterisation of Heavy Metals in Lichen Species *Hypogymnia physodes* and *Evernia prunastri* due to Biomonitoring of Air Pollution in the Vicinity of Copper Mine, *International Journal of Environmental Research*, *6*(3), 779-794.

Balabanova, B., Stafilov, T., Šajn, R. & Bačeva, K. (2014). Comparison of response of moss, lichens and attic dust to geology and atmospheric pollution from copper mine. *International journal of Environmental Science and Technology*, *26*(7), in press, DOI:10.1007/s13762-013-0262-8.

Barandovski, L, Frontasyeva, M. V., Stafilov, T., Šajn, R., Pavlov, S. & Enimiteva, V. (2012). Trends of atmospheric deposition of trace elements in Macedonia studied by the moss biomonitoring technique. *Journal of Environmental Science and Health, Part A, Toxic/Hazardous Substance & Environmental Engineering*, *47*(13), 2000-2015.

Barandovski, L., Cekova, M., Frontasyeva, M. V., Pavlov, S. S., Stafilov, T., Steinnes, E. & Urumov, V. (2006). Air pollution studies in Macedonia using the moss biomonitoring technique, NAA, AAS and GIS Technology; Preprint E18-2006-160, Joint Institute for Nuclear Research: Dubna.

Barandovski, L., Cekova, M., Frontasyeva, M. V., Pavlov, S. S., Stafilov, T., Steinnes, E. & Urumov, V. (2008). Atmospheric deposition of trace element pollutants in Macedonia studied by the moss biomonitoring technique. *Environmental Monitoring and Assessment*, *138*, 107–118.

Barandovski, L., Stafilov, T., Šajn, R., Frontasyeva, M. V. & Bačeva, K., (2013). Air pollution study in Macedonia using a moss biomonitoring technique, ICP-AES and AAS. *Macedonian Journal of Chemistry and Chemical Engineering*, *32*(1), 89-107.

Blagnytė, R. & Paliulis, D. (2010). Research into heavy metals pollution of atmosphere applying moss as bioindicator: a literature review. *Environmental Research, Engineering and Management*, *4*(54), 26-33.

Boev, B. & Jankovic, S. (1996). *Nickel and nickeliferrous iron deposits of the Vardar zone (SE Europe) with particular references to the Rzanovo-Studena Voda ore-bearing series.* Faculty of Mining and Geology, Štip, 103–126.

Boev, B. & Jelenković, R. (2012). *Allchar deposit in Republic of Macedonia – Petrology and Age Determination*, In: Al-Juboury AI (ed) Petrology - New Perspectives and Applications, InTech, Rijeka, Croatia, 131−168.

Box, G. E. P. & Cox, D. R. (1964). An analysis of transformations. *Journal of the Royal Society, Series B (Methodological)*, *26*(2), 211-252.

Ceburnis, U. D. & Valiulis D. (1999). Investigation of absolute metal uptake efficiency from precipitation in moss. *The Science of the Total Environment*, *226*, 247 – 253.

Dołegowska, S., Migaszewski, M. Z. & Michalik, A. (2013). *Hylocomium splendens* (Hedw.) B.S.G. and *Pleurozium schreberi* (Brid.) Mitt.as trace element bioindicators: Statistical comparison of bioaccumulative properties. *Journal of Environmental Sciences*, *25*(2), 340–347.

Fernandez, J. A. Aboal, J. A. & Real, C. (2007). A new moss biomonitoring method for detecting sources of small scale pollution. *Atmospheric Environment*, *41*, 2098-2110.

Fernandez, J. A., Ares, A., Rey-Asensio, A., Carballeira, A. & Aboal, J. R. (2012). Effect of growth on active biomonitoring with terrestrial mosses. *Journal of Atmospheric Chemistry*, *63*(1), 1-11.

Filzmoser, P., Garrett, R. G. & Reimann, C. (2005). Multivariate outlier detection in exploration geochemistry. *Computers & Geosciences*, *31*(5), 579–587.

Frantz, E., Palme, H., Todt, W., El Goresy, A. & Pavićević, M. K. (1994). Geochemistry of Tl–As minerals and host rocks at Allchar (Macedonia). Solar neutrino detection with TI-205 the ''LOREX'' Project; geology, mineralogy and geochemistry of the Allchar deposit locality Crven Dol. *Neues Jahrbuch für Mineralogie Abhandlungen*, *167*, 359–399.

Fränzle, O. (2006). Complex bioindication and environmental stress assessment. *Ecological Indicators*, *6*, 114-136.

Gjengedal, E. & Steinnes, E. (1990). Uptake of metal ions in moss from artificial precipitation. *Environmental Monitoring and Assessment*, *14*, 77–87.

Glukhov, A. Z. & Prokhorova, S. I. (2011). Morphological variability of the brome grass: application for nonspecific biomonitoring of urban anthropogenic impact. *Acta Environmentalica Universitatis Comenianae (Bratislava)*, *19*, 76–81.

Harmens, H., Norris, D. A., Steinnes, E., Kubin, E., Piispanen, J., Alber, R., Aleksiayenak, Y., Blum, O., Coşkun, M., Dam, M., De Temmerman, L., Fernández, J. A., Frolova, M., Frontasyeva, M., González-Miqueo, L., Grodzińska, K., Jeran, Z., Korzekwa, S., Krmar, M., Kvietkus, K., Leblond, S., Liiv, S., Magnússon, S. H., Maňkovská, B., Pesch, R., Rühling, Å., Santamaria, J. M., Schröder, W., Spiric, Z., Suchara, I., Thöni, L., Urumov, V., Yurukova, L. & Zechmeister, H. G. (2010). Mosses as biomonitors of atmospheric heavy metal deposition: Spatial patternsand temporal trends in Europe. *Environmental Pollution*, *158*, 3144-3156.

Harmens, H., Norris, D., Mills, G. & the participants of the moss survey (2013). Heavy metals and nitrogen in mosses: spatial patterns in 2010/2011 and long-term temporal trends in Europe. ICP Vegetation

Programme Coordination Centre, Centre for Ecology and Hydrology, Bangor, UK, 63 pp.

Hock, B. & Seifert, M. (2003). Biomonitoring, a new challenge for measuring and testing. *Management of Environmental Quality: An International Journal*, *14*(2), 279-292.

Hock, B. & Seifert, M. (2003). Biomonitoring, a new challenge for measuring and testing. *Management of Environmental Quality*, *14*, 279-292.

Hristov, S., Karajovanović, M. & Stračkov, M. (1965). *Basic geological map of SFRJ, sheet Štip, M 1:100,000* (map & interpreter), Federal Geological Survey, Beograd.

Janković, S. (1993). Metallogenic features of the Alshar epithermal Sb-As-Tl deposit (the Serbo-Macedonian metallogenic province). *Neues Jahrbuch für Mineralogie Abhandlungen*, *166*(1), 25-41.

Jelenković, R., Boev, B. (2011). Vertical mineralization interval and forecast of the position of an ore-body in the Alšar Sb-As-Tl deposit, Macedonia. *Annales Geologiques De La Peninsule Balkanique*, 72, 119–129.

Jelenković, R. & Boev, B., Pavićević, M. (2011). Determination of primer deepness of ore body of the Sb-As-Tl deposit Allchar (Macedonia), *Tehnika*, *66*(3), 399-405 (in Serbian).

Lazarevski, A. (1993). *Climate in Macedonia.* Kultura: Skopje, (in Macedonian).

Longchurst, J. W. S. & Brebbia, C. A. (2013). *Air Pollution.* Great Britain: WIT Press.

Maksimović, Z. (1982). *Mineralogical Study of the Ore from Ržanovo Mine.* Faculty of Geology and Mining: Belgrade.

Markert, B. A., Breure, A. M. & Zechmeister, H. G. (2003). Definitions, Strategies, and Principles for Bioindication/Biomonitoring of the Environment; Elsevier Press, Oxford, 3 - 39.

Onianwa, P. C. (2001). Monitoring atmospheric metal pollution: a review of the use of mosses as indicators. *Environmental Monitoring and Assessment*, *71*(1), 13-50.

Rakićević, T., Dumurdzanov, N. & Petkovski, M. (1969). *Basic geological map of SFRJ, sheet Štip, M 1:100,000* (map & interpreter), Federal Geological Survey, Beograd.

Rühling, Å. & Steinnes, E., (1998). Atmospheric heavy metal deposition in Europe 1995–1996. NORD 1998: 15, Nordic Council of Ministers, Copenhagen, Denmark.

Rühling, Å. & Tyler, G. (1968). An ecological approach to the lead problem. *Botaniska Notiser*, *122*, 248-342.

Serafimovski, T., Dolenec, T. & Tasev, G., (2006). New data concerning the major ore minerals and sulphosalts from the Pb-Zn Zletovo mine, Macedonia. *RMZ – Materials and Geoenvironment*, *52*, 535–548.

Stafilov, T., Balabanova, B., Šajn, R., Baèeva, K. & Boev B., (2010b) *Geochemical atlas of Radoviš and the environs and the distribution of heavy metals in air*, Faculty of Natural Science and Mathematics, Skopje.

Stafilov, T., Šajn, R., Boev, B., Cvetković, J., Mukaetov, D., Andreevski, M. & Lepitkova, S. (2010a). Distribution of some elements in surface soil over the Kavadarci Region, Republic of Macedonia. *Environmental Earth Sciences*, *61*, 1515-1530.

Stafilov, T., Šajn, R., Pančevski, Z., Boev, B., Frontasyeva, M. V. & Strelkova, L. P. (2010). Heavy metal contamination of surface soils around a lead and zinc smelter in the Republic of Macedonia. *Journal of Hazardous Materials*, *175*, 896–914.

Stafilov, T., Todorovski, T., Grozdanova, B. & Spandževa Lj. (1988), Determination of thallium in ore from Allchar by atomic absorption spectrometry. *Nuclear Instrumenrs and Methods in Physics*, *271*, 321–323.

Stafilov, T., Šajn, R., Boev, B., Cvetković, J., Mukaetov, D. & Andreevski, M. (2008). *Geochemical atlas of Kavadarci and the environs*. Faculty of Science: Sts Cyril and Methodius University, Skopje.

Stefanova, V., Neviealkov, R. & Moritz, R. (2004). *Magmatism of the Borov Dol copper occurrence*, in Kovachev, V. (eds.), *Proceedings on the Annual Scientific Conference of the Bulgarian Geological Society*, Geology House of Science and Technology, Sofia.

Steinnes, E., Rühling, Å., Lippo, H. & Mäkinen, A. (1997). Reference materials for large-scale metal deposition surveys. *Accreditation and Quality Assurance*, *2*, 243–249

Vallero, D. (2008). *Fundamentals of air pollution*, 4th ed., Academic Press Elsevier, New York.

Vrhovnik, P., Šmuc, N. R., Dolenec, T., Serafimovski, T. & Dolenec, M., (2013). Impact of Pb-Zn mining activity on surficial sediments of Lake Kalimanci (FYR Macedonia). *Turkish Journal of Earth Sciences*, *22*, 1-14.

Wolterbeek, B. (2002). Biomonitoring of trace element air pollution: Principles, possibilities and perspectives. *Environmental Pollution*, *120*(1), 11-21.

Zechmeister, H. G., Grodzinska, K. & Szarek-Qukaszewska, G. (2003). 'Bryophytes' in: Markert, B. A., Breure, A. M., Zechmeister, H. G. (Eds.),

Bioindicators and Biomonitors, Elsevier Science Ltd., Amsterdam, 329 - 375.

Zhang, C. S. & Zhang, S. A. (1996). Robust-symmetric mean: A new way of mean calculation for environmental data. *Geo Journal*, *40*, 209-212.

Zhang, C. & Selinus, O. (1998). Statistics and GIS in environmental geochemistry - some problems and solutions. *Journal of Geochemical Exploration*, *64*(1-3), 339-354.

Žibret, G. & Šajn, R. (2008a). Impacts of the mining and smelting activities to the environment-Slovenian case studies. In Sanchez, M.L., Ed.: Causes and effects of heavy metal pollution. Nova Science Publishers, New York; 1-80.

Žibret. G. & Šajn, R. (2008b). Modeling of the atmospheric dispersion of heavy metals in the Celje area, Slovenia. *Journal of the Geochemical Exploitation*, *97*, 29-41.

Žibret. G. & Šajn, R. (2010). Hunting for geochemical associations of elements: factor analysis and self-organizing maps. *Mathematical Geosciences*, *42*, 681-703.

In: Moss
Editor: Jorma Mohamed

ISBN: 978-1-63117-396-7

Chapter 6

BIOMONITORING OF ATMOSPHERIC POLLUTION USING MOSSES: RECENT DEVELOPMENTS AND FUTURE PROSPECTS

Snežana Dragović[1], Mirjana Ćujić[2], Jelena Petrović[2], Ranko Dragović[3], Milan Đorđević[3] and Ljiljana Janković-Mandić[1]

[1]University of Belgrade, Vinča Institute of Nuclear Sciences, Belgrade, Serbia
[2]University of Belgrade, Institute for the Application of Nuclear Energy, Belgrade, Serbia
[3]University of Niš, Faculty of Science and Mathematics, Department of Geography, Niš, Serbia

ABSTRACT

Due to their peculiar morphological and physiological characteristics (lack of a protective cuticle and thick cell walls, numerous cell wall constituents with negatively charged groups, mineral nutrition obtained mainly from wet and dry deposition) mosses are extensively used as biomonitors of environmental contamination for a wide range of contaminants, i.e. radionuclides, heavy metals and toxic nitrogen, sulphur and organic compounds and also as reference organisms in the assessment of radiation exposure to non-human biota. Mosses usually show considerable longevity, which assists their use as long-term integrators of atmospheric deposition. The use of mosses as biomonitors of atmospheric contamination is reviewed in this chapter, with the special reference to terrestrial ecosystems in Serbia, including data on radionuclide and heavy metal content in mosses, their spatial patterns and species- and site-specific dependencies.

INTRODUCTION

Biomonitoring, in the general sense, may be defined as the use of organisms and biomaterials to obtain information on certain characteristics of the biosphere. Organisms, populations, biocenoses and the whole ecosystems are influenced by numerous biotic and abiotic stress factors which affect biological systems at different levels of organisation, from individual enzyme systems through cells, organs, single organisms and population to entire ecosystems. They react not only to single substances or parameters; they show species-specific and situation-specific sensitivity to the number of factors and parameters acting simultaneously at their location. Information on the sensitivity and specificity of such reactions provides a basis for planning the use and evaluating the results of effect-related biological measuring techniques (Markert et al., 2003; Wagner, 1992). Markert et al. (1997) defined biomonitors as "organisms or communities of organisms whose content of certain elements or compounds and/or whose morphological, histological or cellular structure, metabolic-biochemical processes, behaviour or population structure(s), including changes in these parameters, supply information on the quantitative aspects of the quality of the environment or changes in the environment". The relevant information in biomonitoring is commonly deduced either from changes in the behavior of the monitor organism (species composition and/or richness, physiological and/or ecological performance, morphology) or from the concentrations of specific substances in the monitor tissues. To be biologically relevant, a bioindicator should exhibit changes in response to a stressor, have low natural variability, have measurable changes, exhibit persistent changes that are most likely attributable to the stressor, encompass variations in scale and complexity, and embody biologically relevant changes (Burger and Gochfeld, 2001). The changes should be measurable before the damage is irreversible, either to the indicator species itself or to species that depend on it. To be suitable for application as biomonitor of air pollution, specific requirements have to be met by a biological tissue or monitor. According to Sloof (1993), general criteria of primary importance are: a) the response of the organism to the quantity of the pollutants to be monitored should be known; b) the organisms should be common in the area of interest c) availability is required at any time or season; d) the monitor should be tolerant to pollutants at relevant levels. Additional requirements specific for monitoring atmospheric trace elements are: e) the element uptake should be independent of local conditions other than the levels of the elements to be monitored; f) the element uptake should not be influenced by regulating biological mechanisms or antagonistic or synergistic effects; g) the biomonitor should average the element concentrations over a suitable time period as a result of integrated exposure over a time period; h) the organism should not take up appreciable amount of elements from sources other than atmospheric; i) the organisms should have low background concentrations; j) the sampling and sample preparation should be easy and quick; and k) the accumulation should reach to concentration levels which are accessible by routine instrumental measurements. There are some other requirements that should be satisfied in order to obtain the most precise data on the specific pollution source, but they are only partially met. Preferentially, an organism should monitor only one mode of pollution acquirement; in reality, pollutants often reach organisms via more than one route, for example via atmosphere and soil (Rahman and Voigt, 2004).

The advantages of the application of biomonitors compared to the technical equipment are related primarily to the permanent and common occurrence in the field, the ease of sampling and the degree of trace element accumulation. Biomonitors provide a measure of integrated exposure over a certain time, they are present in remote areas and no expensive technical equipment is required for their use (Sloof et al., 1988; Steinnes, 1989).

The Physiological Basis of Mosses for the Use as Biomonitors

Mosses are lower plants without true roots, shoots or vascular system (Zechmeister et al., 2003). They grow in groups forming turfs, cushions or other forms, colonising stones, bark, skeletons, and fresh water. The life history of mosses involves an alternation between sporophytic and gametophytic generations that differ in form and function. The actual plant is represented by the gametophytic generation.

Mosses are poikilohydric organisms, with diversity of means of water and mineral uptake among different species. The most of species are ectohydric, i.e. receive water and mineral nutrients predominantly by atmospheric deposition (Zechmeister et al., 2003). Some species obtain additional water and nutrients from the substrate. Mosses are characterized by desiccation tolerance which is mainly based on physiological adaptations and enables some species to grow even in dry and hot environments. Mineral uptake is controlled by a semipermeable cell membrane. Elements could be associated with moss tissue in different ways: as superficially trapped particulate matter, bound to cell wall on charged exchange sites, intercellular soluble, extracellular or intracellular (Bates, 1992; Brown and Bates, 1990). The total metal binding is species-specific, i.e. it is determined by the number of available exchange sites and morphological structure of moss. Extracellular uptake is rapid and depends on the nature of the elements. Contrary to extracellular uptake, intracellular uptake is influenced by plant metabolism (affinity for certain carrier, competitive elements, gradient in element concentrations). Due to their peculiar morphological and physiological characteristics mosses (division *Bryophytes*) are useful indicators of a wide range of contaminants. Notwithstanding which most bryophytes are small mass and the leaves of many mosses and folious liverworts consist of only one cell layer, the surface-to-volume ratio is pretty high. Bryophytes are resistant against many substances which are often highly toxic for other plants. As a consequence of their nutrient cycling and uptake mechanisms they tend to accumulate pollutants. In terrestrial moss species, all nutrients are mainly absorbed directly from the atmosphere across the cell wall. The accumulation rates of contaminants can be very high and the concentrations found in mosses can reach potentially toxic levels without apparent damage to the plant itself (Poikolainen, 2004).

Mosses as Biomonitors of Atmospheric Pollution in Terrestrial Ecosystems

Radionuclides

The high capacity of mosses to concentrate radionuclides has been known for a long time (Gorham, 1958; Hvinden and Lillegraven, 1961). It has been found that moss *Pleurozium schreberi* absorbed zirconium, boron, barium and lanthanum from fallouts after nuclear tests (Svenson and Linden, 1965). Clymo (1978) and Oldfield et al. (1979) determined the radiocaesium from nuclear weapon tests of the 1960s in the top segments of mosses. Following the Chernobyl reactor accident numerous studies involving monitoring of the long-lived fission product ^{137}Cs have been undertaken using mosses as biomonitors (Elstner et al., 1987; Giovani, 1994; Nifontova, 1995; Papastefanou et al., 1989; Steinnes and Njåstad, 1993). Mosses have also been widely used for the assessment of natural (Belevermiş and Çotuk, 2010; Boryło et al., 2012; Dragović et al., 2007; Popic et al., 2011) and technologically-enhanced radioactivity (Abu Bakar et al., 2013; Delfanti et al., 1999; Dragović and Nedić, 2010; Uğur et al., 2003;). They found to be useful biomonitors of atmospheric deposition of cosmogenic ^{7}Be (Dragović et al., 2007; Krmar et al., 2007; Krmar et al., 2013). In Serbia, mosses are used for the assessment of natural and anthropogenic radioactivity in terrestrial ecosystems (Dragović et al., 2004; Grdović et al., 2010; Popović et al., 2008; Popović et al., 2009). These studies have shown considerable variations in radionuclide uptake among moss species and also among sampling locations. The highest activity concentrations of ^{137}Cs have been found in ecosystems at higher altitudes predisposed to receive higher fallout because of high precipitation rates, which enhance deposition. Since these ecosystems present almost closed biogeocenoses, with efficient nutrient cycling between its components, the level of radioactive contamination remains high for a long time. The variations in ^{137}Cs activity concentrations have found to exist even in mosses from the small-scale areas (Dragović et al., 2010b; Figure 1). The transfer of radionuclides in mosses has been quantified by concentration ratios as they are reference organisms in the assessments of radiation exposure of non-human biota (Dragović et al., 2010a, 2010b; Dragović and Janković-Mandić, 2010). The relationship between concentration ratios and physical and chemical characteristics of the underlying soil has also been investigated (Dragović et al., 2010b).

Even though mosses were extensively used and studied as biomonitors, the mechanisms of trace element uptake, retention and tolerance are still not sufficiently known. These organisms exhibit a range of reactions at a cellular level that might be involved in detoxification and, thus, tolerance to a stress caused by the environmental pollution (Hall, 2002). Experimental investigation of ^{137}Cs binding strength in mosses, as well as the specificity of binding, has been conducted using aqueous solutions of salts and acids. The efficiency of different cations to exchange with $^{137}Cs^{+}$ in mosses has been investigated (Dragović et al., 2002).

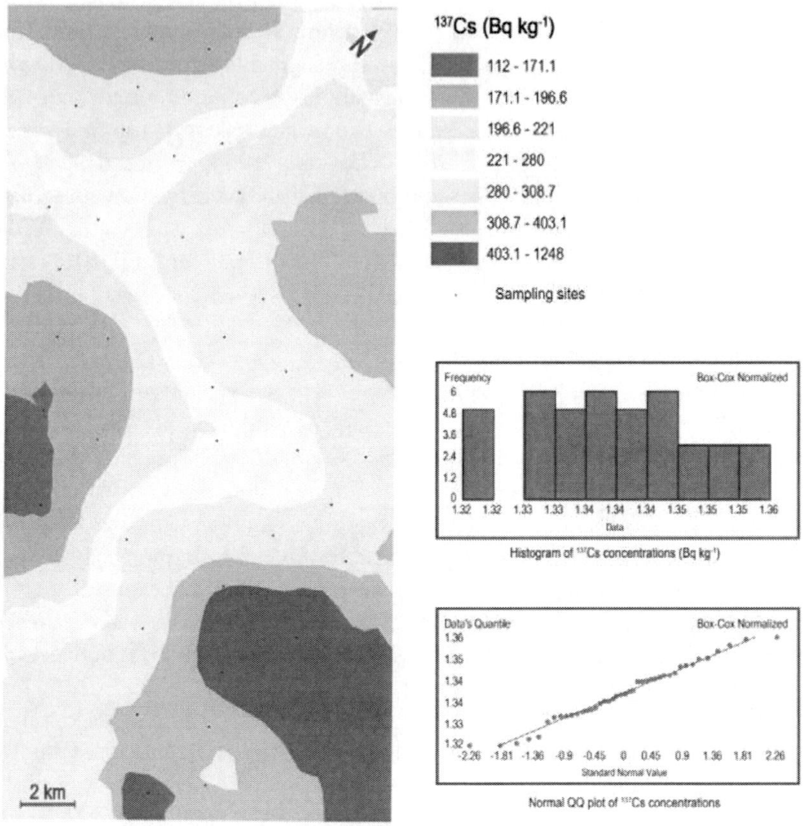

Figure 1. Spatial distribution of ^{137}Cs activity concentrations in mosses from Zlatibor Mountain.

Ammonium cation proved to be the most suitable cation to substitute $^{137}Cs^{+}$ in moss, followed by sodium, potassium and caesium. In the series of different ammonium salts, ammonium phosphate solution was the most effective for $^{137}Cs^{+}$ displacement, followed by ammonium chloride, thiocyanate and sulphate. As for the efficiency of organic salts, ammonium oxalate reduced the presence of $^{137}Cs^{+}$ in the analyzed organisms to the greatest extent, followed by other oxalates and ammonium acetate (Nedić and Dragović, 2006). The ability to exchange $^{137}Cs^{+}$ for H^{+} in mosses has been further tested, using different inorganic acids (hydrochloric, sulphuric, phosphoric and boric). All inorganic acids, except boric, were able to significantly decrease the amount of $^{137}Cs^{+}$ in mosses (Dragović et al., 2002). Using phosphoric acid 30% more $^{137}Cs^{+}$ has been extracted than in distilled water. The chemical behaviour of caesium is similar to that of potassium and structures that bind Cs^{+} in a cell are those with the greatest affinity for K^{+}. Physiological studies on higher plants demonstrated that Cs^{+} and K^{+} competed for influx, suggesting that the influx of these cations is mediated by the same mechanism(s) (White and Broadley, 2000; Zhu and Smolders, 2000). The studies on accumulation of ^{137}Cs in the algae *Nitella flexilis* and *Chara coralline* have shown that caesium is accumulated in the cell wall and then enters a cell through potassium channels in the plasma membrane (Demidchik et al., 2001; Draber and Hansen, 1994). Potassium channels are also permeable to other monovalent cations, like Na^{+}, Li^{+}, Rb^{+}, NH_4^{+}, Tl^{+}, as their size (0.19-0.30 nm) is similar to that of K^{+} (0.27 nm). Potassium channels, however, conduct these cations at different rates (Choe et al., 2000). Besides the radius of the hydrated ion, the three dimensional structure of the binding sites is of the greatest importance. The selectivity in ion conduction is based on different affinities of the binding sites for cations and different rates of their translocation. Although a general trend of the ion exchange efficiency was kept, examined moss species exhibited different behaviour, in quantitative terms, upon subjection to salt and acid solutions (Dragović et al., 2002). Different species of the same type, within moss classification, collected at the same time and location, had significantly different specific radioactivity and desorption rates of ^{137}Cs. It is, therefore, obvious that biodiversity is a variable that must be taken into consideration when discussing the type and strength of $^{137}Cs^{+}$ binding.

The investigations focused on closer determination of the relation between organic molecules and $^{137}Cs^{+}$ in mosses (Dragović et al., 2004; Hranisavljević-Jakovljević et al., 1980) have shown that almost all ^{137}Cs (93-94% of the initial activity) can be extracted using neutral and acidic aqueous solutions without complete destruction of the moss tissue. Fractional precipitation of aqueous extracts with ethanol indicated that essential biomolecules, polypeptides and polysaccharides, do not bind significant amounts of ^{137}Cs (Dragović et al., 2004). The great tolerance of mosses towards radionuclides and other toxic substances could be partly explained by this finding.

For better understanding of the high tolerance of mosses to radiocaesium and other toxic substances the distribution and cellular compartmentation of ^{137}Cs has been investigated (Dragović et al., 2004). The ability of a cell wall and plasma membrane to bind and, in a way, "immobilize" ^{137}Cs, thus avoiding translocation into intracellular space, has been investigated using three different moss samples. The cellular fractionation was performed by differential centrifugation of tissue homogenates obtained from (Becker et al., 1995). The presence of ^{137}Cs was monitored by measuring the radioactivity level in each fraction. The greatest amount of ^{137}Cs was found in fractions containing plasma membranes, ribosomes, endoplasmatic reticulum and microsomes, which precipitated after ultracentrifugation at moss tissues have shown that cell walls contain 25-30% of the initial ^{137}Cs in moss, plastids, chromoplasts and mitochondria 17-25% of the initial ^{137}Cs and the remaining cytosol had 10-13%. A small quantity of ^{137}Cs in the soluble fraction of the cell lysate, thus, indicates that relatively small amount of the radiocaesium is present as a free cation in a cell. The cell wall also has found to be an efficient barrier against the penetration of heavy metals into the protoplasm of the moss cell (Shimwell and Laurie, 1972; Skaar et al., 1973). Burton (1979) found that *Fontinalis antipyretica* maintained 80-90% of its accumulated Zn in the cell walls. The cellular fractionation demonstrated that the relative distribution of ^{137}Cs between subcellular fractions is genus independent. This finding pointed out the similar chemical structure of ^{137}Cs-binding molecules in different mosses. Their distribution between various cellular compartments is not found to be species specific.

Heavy Metals

Heavy metals can be deposited on terrestrial surfaces and subsequently build up in soil or sediment, they are persistent in the environment and may be concentrated in bioaccumulator species like mosses. The moss biomonitoring technique for measurement of atmospheric heavy metal deposition was developed in the late 1960. Idea of using mosses for this purpose is based on the fact that carpet forming ectohydric mosses obtained most trace elements and nutrients directly from precipitation and dry deposition and that uptake of metals from the substrate is insignificant (Rühling and Tyler, 1968, 1970; Tyler, 1970). Carpet-forming terrestrial mosses (*Hylocomium splendens*, *Pleurozium schreberi*, *Hypnum cuppressiforme*, *Scleropodium purum,* etc.) are widely used to determine and monitor atmospheric pollution by heavy metals. Mosses show some advantages as biomonitors of heavy metals: they are perennial without deciduous periods; they have a high cation-exchange capacity that allows them to accumulate a large amounts of heavy metals between the apoplast and symplast compartments without damaging vital functions of the cells (Harmens et al., 2013); the presence of polygalacturonic acids on the external side of the cell wall and proteins in the plasma membrane influence cation-exchange capacity (Aceto et al., 2003); mosses do not possess thick, strong protective layers like cuticles. The characteristic physiology and morphology of mosses imply bioconcentration and passive uptake of contaminants from the atmosphere. Advantage of using mosses to measure metal deposition from the atmosphere to terrestrial ecosystem is lying in the fact that it is easier and cheaper than conventional precipitation analysis and much higher sampling density can be achieved using mosses.

In the last two decades mosses have been used to monitor and map the atmospheric deposition of major, minor and trace element in large scale surveys in Europe (Aceto et al., 2003; Gombert et al., 2004; Grodzińska and Szarek-Łukaszewska, 2001; Harmens et al., 2008; Lucaciu et al., 2004; Ötvös et al., 2003; Sucharová and Suchara, 2004). The first moss survey at the European scale was conducted in 1990 and is now carried out at five-year intervals, since then the number of participating countries has greatly expanded. The ICP Vegetation survey provides data on concentrations of ten heavy metals (arsenic, cadmium, chromium, copper, iron, lead, mercury, nickel, vanadium and zinc) in naturally-growing mosses throughout Europe (Harmens et al., 2013). Since 2005, data on aluminium, antimony and nitrogen are also included. Based on ICP vegetation reports deposition of heavy metals across Europe in recent decades has resulted in a decrease in the heavy metal concentration in mosses since 1990. The metal concentration in mosses between 1990 and 2010 has declined the most for lead (77%) and the least for copper (6%). Between 1995 and 2010, the average mercury concentration in mosses has declined by 23%. The decreased concentrations of heavy metals could be attributed to the fact that during the years many pollution sources reduced their emissions and also some very polluting local emission sources have been shut down since 1990, particularly in Eastern Europe. When comparing changes, it should bear in mind that there is possibility that sampled mosses are not collected from the same sites each year and/or the sampling density or strategy between years might have changed. Also the application of different analytical techniques and/or improvements in the performance of analytical instruments might have contributed to the variation in element concentrations in mosses between years. The lowest concentrations of heavy metals in mosses were generally found in northern Europe, although higher concentrations were reported near local sources. For western and central Europe low to intermediate heavy metal concentrations in mosses were reported, while the highest concentrations were found for (south-) eastern

Europe (Harmens et al., 2013). Elevated concentrations of heavy metals in mosses can arise in several ways: contemporary or historical industrial and mining activities; large conurbations; vehicle emissions along major roads; geological sources, or long-range transport of pollution.

To evaluate the behaviour of some moss species as bioaccumulators, to determine the maximum level of an element potentially available to a moss from the soil, and also to evaluate the contribution to the metal content in moss from anthropogenic sources, enrichment factors (EFs) could be calculated. Applied to mosses, enrichment factors compare the concentration of analyte of interest to the concentration of a conservative soil element (generally aluminium, iron, silica or titanium) in moss versus local parent material (Puxbaum, 1991). Ratio values above 10.0 are generally considered to reflect plant contaminant concentrations in excess of what would normally be supplied by the local geological substrate.

Element deposition using mosses in the 2000 French survey aimed to identifying sources and spatial trends of investigated elements (Gombert et al., 2004). In that study, elements with lowest concentrations in mosses are considered to be toxic for plants: Pb, Cd and Hg, with median concentrations of 5.7, 0.2 and 0.3 mg/kg respectively. For those elements, a significant share of the emissions derives from anthropogenic sources, mainly from combustion processes. Coefficients of variation higher than 100% were found for concentrations of Al, Cr, Fe and Mn. According to Reimann et al. (2001) large element variability provides an indication of additional element source, which can be related to an atmospheric pollution for some elements. In moss studies the range of chemometric methods are applied to concentration data to find out relevant information (Aceto et al., 2003; Gombert et al., 2004; Viet et al., 2010). They are often combined with enrichment factor analysis to reveal the origin of the elements of interest in mosses (Gombert et al., 2004). Survey carried in Hungary using mosses pointed out that one of the most polluted locations with high concentrations of Cd, Cr, Fe, Ni, V was Tiszaújváros, where a chemical works can be found, evaluated concentrations of Cd, Fe, Ni and V were found near an oil refinery and power plants in Százhalombatta and near Budapest all investigated heavy metals were mildly elevated (Ötvös et al., 2003). When mosses are used for long scale survey investigation, different aims can be achieved. The study of Aceto et al. (2003) aimed to verify whether a correlation exists between metal content in moss and pollution state of the site where moss is sampled. The comparison of concentrations with pluviometric and thermal data showed the strict dependence of metal content in moss on meteorological conditions. Influence of sampling season was investigated by Fernandez and Carballeira (2002) and revealed that when large scale studies are carried out in Galicia using the species *S. purum*, the sampling does not have to be carried out at any particular time of year. For *Polytrichum formosum* significant variation in the heavy metal concentrations during the year was found. This could be explained by differentiated cell structure, which can protect the leaf surface and facilitates uptake of water and nutrients from the soil (Markert and Weckert, 1989).

A number of studies resulted in maps of the regional depositions patterns of heavy metals based on moss monitoring data (Gombert et al., 2004; Grodzińska and Szarek-Łukaszewska, 2001; Lucaciu et al., 2004; Ötvös et al., 2003; Suchara and Sucharová, 2004).

Mosses have also been widely used as biomonitors in smaller-scale surveys to estimate the deposition of pollutants in industrial zones (Bargagli et al., 2002; Culicov et al., 2002; Ćujić et al., 2013; Fernández and Carballeira, 2002; Gondzáles-Miqueo et al., 2010; Goodarzi and Senei, 2005; Grodzińska et al., 2003; Uğur et al., 2004; Viet et al., 2010). Overview of heavy metal concentrations derived in those studies is presented in Table 1.

Table 1. Overview of mean heavy metal concentrations (mg kg^{-1}) in mosses obtained in industrial areas worldwide

Element	Poland	Romania	Serbia	Spain	Canada	Turkey	Vietnam
	Grodzińska et al., 2003	Culicov et al., 2002	Ćujić et al., 2013	Fernández and Carballeira, 2002	Goodarzi and Senei, 2005	Uğur et al., 2004	Viet et al., 2010
As	0.43	2.17	0.77	0.45	4.1	-[a]	9.9
Cd	2.17	1.43	1.45	0.17	1.76	0.86	1.1
Co	0.3	5.91	4.82	0.52	5.1	7.2	1.5
Cr	6.2	1.48	10.95	1.42	-	14.61	16
Cu	9.4	33	21.97	7.4	19.8	12.71	17.7
Fe	1226	1032	2315	583	12700	4884	4700
Hg	-	-	2.65	0.09	0.04	-	-
Mn	145	222	127	285	488	148	153
Ni	2.5	1.96	17.32	2.0	14.1	22.82	4.9
Pb	62	46	2.91	5.4	10	37.1	74.9
V	3.9	-	12.69	-	-	-	53
Zn	150	-	32.58	58	97	145.14	121

[a]Data not reported.

The conventional view is that moss chemistry reflects atmospheric element inputs, but other sources of inputs for some elements can influence moss composition. In many studies sources and the types of pollutants to which terrestrial mosses are exposed were analysed.

Principal component analysis is useful tool to distinguish different types of element inputs, and in general there are three of them: atmospheric, edaphic and biotic. Atmospheric sources include long range atmospheric transport or local point sources of metals and metalloids such as: As, Bi, Cd, Cu, Hg, Mo, Ni, Pb, Sb, Sn, Tl, V and Zn. Atmospheric inputs also include natural sources such as sea salts and biogenic emissions from the marine environment: B, Br, Ca, Cl, I, Li, Mg, Na, S, Se, Sr and U. Elements derived in forest fire and in volcanic emission contribute to atmospheric sources (Reimann et al., 2001). The edaphic inputs mainly correspond to mineral particles from the soil carried by the wind and deposited into the moss tissues. These inputs also may include substrate influence due to human activities related to mining activities. The elements mainly related to the edaphic inputs are Al, Co, Cr and Fe. Aluminium is present at high concentrations in the Earth's crust and is a good indicator of mineral particles, mainly windblown soil dust (Berg and Steinnes, 1997; Zechmeister et al., 2003). Iron is a major element in bedrock and soil and is an essential micronutrient for all organisms. The concentrations of 800 mg kg^{-1} of iron have been shown to cause morphological damage in sensitive *Peltigera spp.* (Goyal and Seaward, 1982). Correlations between metal concentrations in moss and the surrounding vegetation indicate that elements such as Mn and Zn, could be subsequently transferred to mosses by leaching from plant material (Aboal et al., 2004; Čerbunis and Steinnes, 2000). Fernandeź et al. (2004) were found that mosses could accumulate different concentrations of contaminant depending on whether it is metal or metalloid. Processes related to the cation-exchange capacity of the cell membrane will determine the final concentration in moss.

In Serbia, mosses have been used for the assessment of atmospheric pollution by heavy metals at different spatial scales. Concentration ranges obtained for investigated heavy metals for different parts of Serbia are presented in Table 2. In the study of Frontasyeva et al. (2004) mosses *Hypnum cupressiforme* were collected from the northern part of Serbia and the obtained data subjected to principal component factor analysis to reveal the origin of elements. High concentrations of vanadium in the moss samples at certain areas have been suspected to be associated with fires in oil refineries damaged during the 1999 military action. Dragović and Mihailović (2009) have investigated the origin of heavy metal pollution in the Zlatibor ecosystem using mosses. Different moss species (*Pleurozium schreberi, Hylocomium splendens, Scleropodium purum, Hypnum cupressiforme* and *Thuidum delicatulum*) and underlying soils (0-5 cm) have been analyzed and EFs of seven heavy metals calculated. The EFs varied by a factor of 38 among the elements analysed. Mosses from the investigated area have been enriched in Cd and Pb from anthropogenic activities and Mn by leaching from higher plants. The EFs of Cd, Mn and Cr were 22.5, 60.4 and 25.2, respectively. The pollution in this area remote from local industrial facilities has been probably caused by both long-range transport and fossil fuel burning. Mosses have also been used to assess atmospheric deposition in the vicinity of the largest thermoelectric power plant in Serbia "Nikola Tesla" in the municipality Obrenovac. The concentrations of 15 elements in different moss species (*Bryum argenteum, Bryum capillare, Brachythecium sp.* and *Hypnum cupressiforme*) have been analysed and geographically mapped to reveal the areas of the highest contamination (Sabovljević et al., 2005, 2007, 2009; Vukojević et al., 2006, 2009). Ćujić et al. (2013) have conducted the assessment study in the same area (genus *Brachythecium sp.* and specie *Kindbergia praelonga* (Hedw.) Ochyra), determined the concentration of 22 elements (Al, As, B, Ba, Ca, Cd, Co, Cr, Cu, Fe, Hg, K, Mg, Mn, Na, Ni, Pb, Si, Sr, Ti, V, Zn) in moss samples, analysed their mutual relationships and distinguished their lithogenic or anthropogenic origin by cluster analysis. Three clusters of elements were noticed: the first comprised Fe, Al, Ba, Ti, Cr, Mn, As and Na; the second one consisted of K, Si, Zn, Mg, Ca, Sr and Ni; while Co, B, Pb, Hg, Cu, Cd and V were grouped in the third cluster. The rare crustal elements As, Cd, Co and Hg, as well as Ni and V showed elevated concentrations in mosses in the vicinity of thermal power plants. Mean values for the elements investigated have found to be lower than those found in the same region in previous studies (Sabovljević et al., 2005, 2007, 2009; Vukojević et al., 2006, 2009) due to implementation of some countermeasures, such as the replacement of six electrostatic precipitators, modernisation of ash handling and the disposal system to reduce air, ground and underground water pollution, compilation of a database regarding terrain stability in the thermal power plant zones and improvement of the mechanism for waste management. In an active biomonitoring study, in the urban area of Belgrade, accumulation of trace elements (Al, V, Cr, Mn, Fe, Ni, Cu, Zn, As, Cd, Pb) in the moss bags (*Sphagnum girgensohnii* Russow) and their relationship to the atmospheric bulk deposition measurements have been investigated (Aničić et al., 2009a). Seasonal variations have been detected in both dry and wet moss bag samples for all of the elements except Pb, Al, and Mn. The highest variations have been noticed for V and Ni, whose content was two and three times higher in winter than in summer, respectively (Aničić et al., 2009a). To assess the element accumulation in moss, the relative accumulation factors (RAF) have been calculated as the ratio of the moss content of element after and before the exposure: $(C_{exposed} - C_{initial})/C_{initial}$.

Table 2. Summary of mean heavy metal concentrations (mg kg^{-1}) in mosses from different areas in Serbia

Element	Northern Serbia[a]	Obrenovac[a]	Zlatibor ecosystem[a]	Belgrade[b]
	Frontasyeva et al., 2004	Ćujić et al., 2013	Dragović and Mihailović, 2009	Aničić et al., 2009[a,b]
As	0.46 - 61	0.47 - 1.77	-[c]	0.38 - 2.2
Cd	0.4 - 6.5	0.62 - 3.49	0.01 - 4.20	0.19 - 0.36
Co	1.42 - 39	0.10 - 10.90	-	0.52 - 1.6
Cr	1.14 - 22	1.70 - 19.70	0.01 - 74.9	2.0 - 6.8
Cu	6.31- 3140	14.27 - 37.23	2.00 - 27.3	10 - 49
Fe	720 - 9230	517 - 5360	-	997 - 1.830
Hg	-	1.53 - 3.49	-	-
Mn	30 - 2340	54 - 201	42.4 - 1587	221 - 239
Ni	1.96 - 26	8.64 – 25.97	0.01 - 182	5.1 - 9.9
Pb	-	0.84 - 26.57	0.01 - 98.1	7.0 - 38
V	2.85 - 39	4.90 - 25.20	-	2.9 - 112
Zn	14 - 415	14.70 - 78.43	21.6 - 65.2	44 - 105

[a]Native moss, [b]Moss bags, [c]Data not reported.

The most accumulated elements in dry moss bags, according to the RAF value have found to be as follows: V (22) > Cr (11) > Cu (9) > Pb (8) > As (5) > Al (4) > Fe (3) > Ni (3) ≈ Zn (2.5) > Mn (0.9) > Cd (0.5). In wet moss bags, the order for the most accumulated elements was somewhat different: Cu (68) > V (26) > Cr (21) > Pb (13) > Al (6.5) > As (6) > Fe (5) > Zn (4.5) > Ni (4) > Cd (1) > Mn (0.2) (Aničić et al., 2009b).

Nitrogen Compounds

In the atmosphere the nitrogen compounds are mainly present in the oxidised or reduced forms of nitrogen. Many different natural and anthropogenic sources are responsible for emissions of nitrogen compounds to the atmosphere such as: lighting, volcanoes, biological processes in the soil and in aquatic ecosystems, forest fires, combustion processes in energy production, industry and transport, spreading and production of animal manure, application and production of fertilizer (Erisman et al., 1998; Harmens et al., 2011; Oliver et al., 1998).

Table 3. The total nitrogen concentration (% dry weight) in mosses from 16 European countries (Harmens et al., 2011)

Country (region)	Mean (Min-Max)
Austria	1.21 (0.70-2.15)
Belgium	1.49 (0.79-2.31)
Bulgaria	1.45 (0.34-2.94)
Czech Republic	1.16 (0.68-2.30)
Estonia	0.96 (0.63-1.66)
Finland	0.85 (0.38-1.79)
France	1.29 (0.62-2.18)
Germany	1.54 (0.78-3.36)
Italy (Bolzano)	1.12 (0.84-1.52)
Latvia	1.11 (0.80-1.65)
Slovakia	1.80 (0.90-3.82)
Slovenia	1.83 (0.82-2.82)
Spain (Galicia and Navarra)	1.10 (0.61-2.30)
Switzerland	1.20 (0.78-2.12)
Turkey (Thrace)	1.40 (0.78-2.01)
United Kingdom	0.89 (0.44-2.45)

A number of studies have indicated the stability of mosses as indicators of atmospheric nitrogen deposition (Leith et al., 2005; Pitcairn et al., 1995; Zechmeister et al., 2008). Salemaa et al. (2008) suggested that *P. schreberi, H. splendens* and *Dicranum polysetum* can be used as accumulator indicators of nitrogen deposition based on determined linear relationship between the nitrogen concentration of bryophyte shoots and the amounts of given nitrogen. Solga et al. (2005) have found significant relationships between bulk nitrogen deposition and tissue N concentration of *P. schreberi* and *S. purum*. Some studies have shown that species-specific responses should be taken into account (individual species responses of nitrogen deposition) when bryophytes are used as biomonitors of nitrogen deposition (Harmens et al., 2011; Salemaa et al., 2008). Table 3 summarizes data on total nitrogen concentration in mosses (*P. schreberi, H. splendens, H. cupressiforme, Pseudoscleropodium purum* and other) for 16 European countries reported by Harmens et al. (2011). The lowest total nitrogen concentration have been found in mosses from northern Finland and northern parts of UK, while the highest ones in the parts of Belgium, France, Germany, Slovakia, Slovenia and Bulgaria.

The influence of nitrogen deposition on changes in bryophyte communities was explained by correlations between increased deposition and increasing or decreasing species abundance, increased N concentrations or declined nitrate reductase activity of plant tissues (Hallingbäck, 1992; Pearce and Van der Val, 2002; Pitcairn et al., 1995; Solga et al., 2005).

Different factors affect the nitrogen concentration in mosses. Pitcairn et al. (2006) have found that nitrogen content of moss tissue depends of ways (wet or dry) of nitrogen deposition, the tissue nitrogen concentration is higher at dry deposited sites than at wet deposition sites, and also found that tissue nitrogen responds less to the nitrogen deposition than to rainfall concentration of nitrogen (NO_3^- and NH_4^+) at wet deposition sites. The deposition of nitrogen in the atmosphere increases with altitude as a consequence of precipitation increase, frequent cover by orographic cloud and orographically enhanced deposition by "seeder-feeder" effects (Fowler et al., 1998). However, the relationship between the total nitrogen concentration in mosses and altitude was not found to be distinct. Hicks et al. (2000) reported that foliar nitrogen concentrations increased linearly with annual total nitrogen deposition, but did not found significant relationship between foliar nitrogen concentration and altitude for *H. splendens*. Zechmeister et al. (2008) reported significant correlation between altitude and nitrogen content in mosses tissues. Ayres et al. (2006) indicated that moss can uptake the soil nitrogen and have shown that two physiological and taxonomic different mosses species *Racomitrium lanuginosum* and *Polytrichum alpinum* have the capacity to take up nitrogen from the soil.

Sulphur Compounds

The main anthropogenic and natural sources of sulphur emissions are: combustion of fossil fuels, biomass burning, oceans, wetlands, plants and soils, volcanoes etc. In the atmosphere sulphur occurs in different gaseous, liquid and solid chemical forms; the major species present in the atmosphere are: hydrogen sulphide, dimethylsulphide, carbon disulphide, carbonyl sulphide and sulphur dioxide. They are removed from the atmosphere by wet and dry deposition.

Xiao et al. (2011) reported mean value of 0.29% (0.18-0.52%) for tissue sulphur contents of *Haplocladium microphyllum* in Nanchang region (SE China). They found higher sulphur content of urban mosses than of rural mosses, as well as higher sulphur concentration in rainwater at urban than at rural areas.

Liu et al. (2009) reported range from 0.25 to 0.56% for sulphur content of mosses (epilithic and terricolous) under different growing conditions (open fields and different canopy conditions). The higher value of tissue sulphur in the open field have been found in terricolous (confronted with higher sulphur supply due to contribution from subsoil) mosses compared with epilithic mosses. The same study revealed the higher sulphur contents in mosses under canopies than in those from open fields.

Some studies have demonstrated the influence of sulphur pollutants on the growth of many *Sphagnum* species (Ferguson et al., 1978; Ferguson and Lee, 1980).

Nriagu and Glooschenko (1992) found that sulphur content of *Sphagnum fuscum* decreases with distance from smelter at Sudbury, Canada, from 800 $\mu g\ g^{-1}$ at to less than 600 $\mu g\ g^{-1}$ as well as that near Rouyn-Noranda smelter were very few occurrences of *S. fuscum*.

Raymond et al. (2010) reported mean tissue sulphur content of 0.099% (range from 0.07 to 0.16%) for *Isothecium myosuroides* in southwest British Columbia.

Szarek and Chrzanowska (1991) demonstrated no significant difference between the total sulphur content in mosses collected in 1975 and 1986 from national parks in Poland, and also indicated that mosses are not stable indicators of atmospheric pollution by sulphur compounds.

Organic Compounds

Persistent organic pollutants (POPs) originate mostly from anthropogenic sources such as: fossil fuel burning, waste incineration and industrial production and application. POPs are accumulating in the environment due to weak degradability. Many studies have used mosses as bioindicators of atmospheric deposition of POPs, mainly polycyclic aromatic hydrocarbons (PAHs) (Carballeira et al., 2006; Dołęgowska and Migaszewski, 2011; Gałuszka, 2007; Migaszewski et al., 2009; Ötvös et al., 2004).

Table 4. ATSDR/USEPA priority PAHs, and their phase distribution (Ravindra et al., 2008)

PAH[a]	Particle/gas phase distribution
Acenaphthene	Gas phase
Acenaphthylene	Gas phase
Anthracene	Particle gas phase
Phenanthrene	Particle gas phase
Pyrene	Particle gas phase
Benz[a]anthracene	Particle phase
Chrysene	Particle phase
Benzo[b]fluoranthene	Particle phase
Benzo[j]fluoranthene	Particle phase
Benzo[k]fluoranthene	Particle phase
Benzo[a]pyrene	Particle phase
Benzo[e]pyrene	Particle phase
Fluoranthene	Particle gas phase
Fluorene	Gas phase
Dibenz[a,h]anthracene	Particle phase
Benzo[ghi]perylene	Particle phase
Indeno[1,2,3-c,d]pyrene	Particle phase

[a]Predominant sources are motor vehicles and wood smoke.

The sources of PAHs in the environment can be anthropogenic and natural such as: incomplete combustion of organic material, volcanic eruptions, forest fires etc. (Ravindra et al., 2008).

The priority PAHs, emission sources and phase distribution (ATSDR, 1995) are shown in Table 4. Excluding benzo[j]fluoranthene and benzo[e]pyrene from those presented in Table 4 and including naphthalene they are known as 16 priority PAHs for monitoring at USEPA (United States Environmental Protection Agency) pollutant list (USEPA, 1997).

Pleurocarpous mosses (*Abietinella abietina, H. splendens, H. cupressiforme, P. schreberi, P. purum, Sanionia uncinata, S. purum, Thamnobryum alopecurum* and *T. tamariscinum*) and acrocarpous mosse (*Dicranum scoparium* and *Tortula muralis*) have been used as passive biomonitors of atmospheric PAH deposition (Cabrerizo et al., 2012;

Dołęgowska and Migaszewski, 2011; Foan et al., 2010; Foan and Simon, 2012; Migaszewski et al., 2009; Knulst et al., 1995; Krommer et al., 2007; Milukaite, 1998).

After emission to the atmosphere in the gaseous phase remain volatile PAHs which can transported from main pollution sources, non-volatile ones are adsorbed onto plant species and/or solid particles and deposited near emission sources, and medium volatile PAHs are distributed between adsorption and gas phases (Thomas, 1986).

Studies have shown that PAHs with smaller ring numbers are dominant in mosses located away from pollution sources (Dołęgowska and Migaszewski, 2011; Gałuszka, 2007; Migaszewski et al., 2009; Orliński, 2002). The biomonitoring study, which included moss samples *H. cupressiforme* from clean sites (distant from main and minor roads, villages and buildings) in Hungary, have demonstrated that 99% of the total PAHs atmospheric deposition were low molecular weight PAHs (Ötvös et al., 2004).

Ares et al. (2009) have found that the concentrations of PAHs in moss *P. purum* is related to the distance from the emission source and decreases rapidly with increasing distance from source.

Ötvös et al. (2004) have not found correlation between concentration of total PAHs and metal concentrations in *H. cupressiforme*. Gałuszka (2007) also demonstrated no significant correlation between element concentration and PAHs. These findings could be attributed to different emission sources and atmospheric transport pathways of organic pollutants and metals.

The PAHs concentration is mainly influenced by domestic emission and vehicular of the area. Gerdol et al. (2002) reported that total PAHs concentration was lower in rural then in urban area in Northern Italy. Ötvös et al. (2004) reported positive correlation between traffic volume and total PAHs concentration in *H. cupressiforme*.

The bioaccumulations of PAHs in mosses can be species-specific. Dołęgowska and Migaszewski (2011) and Gałusyka (2007) reported that *H. splendens* is more efficient accumulator of PAHs than *P. schreberi*.

Conclusion and Future Prospects

Biomonitoring using mosses have brought about numerous developments in method that have made it possible to meet the demands for protection of the environment against an increasing number of environmental stressors.

The accumulative and time-integrative behaviour of mosses and the relative easy sampling and analysis make biomonitoring using these organisms foreseeable in future surveys of atmospheric pollution. Results of monitoring of atmospheric pollution using mosses have shown that the activity levels found depend on many factors, such as specific strain reaction of species selected, their age, climatic conditions of deposition and procedures employed for their analysis. In order to obtain reliable quantification of transfer of contaminants to biota much effort has devoted to harmonization of sampling protocols and analysis methods. There is still a need for redefinition or rescaling of existing biomonitoring tools which are typically developed for small geographical areas to make them applicable to other areas. They can yield valid data only if designed in compliance with geostatistical requirements and criteria of representativeness.

The knowledge of contaminant transfer to biota should be completed by a better understanding of bioaccumulation of contaminants in living organisms at tissue, cellular and molecular levels, the effects of bioaccumulation at low chronic doses and their characteristics at ecosystem level.

The biomonitoring should in future follow a comprehensive approach, i.e. in addition to determining effects at the level of individual it should enable conclusions at the ecosystem level. It should also ensure the comparability of effects determined in laboratory and in the field.

Acknowledgments

This work was supported by the Ministry of Education, Science and Technological Development of the Republic of Serbia (Project No. III43009). The authors are grateful to Dr Mira Aničić-Urošević, Institute of Physics, University of Belgrade, who reviewed this chapter, for her careful reading of the text and helpful remarks.

References

Aboal, J. R.; Fernandeź, J. A.; Carballeira, A. *Environ. Exp. Bot.,* 2004, 51, 215-225.

Abu Bakar, N. S.; Wan Mahmood, Z. U.; Saat, A. *J. Radioanal. Nucl. Chem.,* 2013, 295, 315-323.

Aceto, M.; Abollino, O.; Conca, R.; Malandrino, M.; Mentasti, E.; Sarzanini, C. *Chemosphere,* 2003, 50, 333-342.

Aničić, M.; Tasić, M.; Frontasyeva, M. V.; Tomašević, M.; Rajšić, S.; Mijić, Z.; Popović, A. *Environ. Pollut.,* 2009a, 157, 673-679.

Aničić, M.; Tasić, M.; Frontasyeva, M. V.; Tomašević, M.; Rajšić, S.; Strelkova, L. P.; Popović, A.; Steinnes, E. *Environ. Chem. Lett.,* 2009b, 7, 55-60.

Ares, A.; Aboal, J. R.; Fernández, J. A.; Real, C.; Carballeira, A. *Atm. Environ.,* 2009, 43, 5501-5509.

ATSDR (Agency for Toxic Substances and Disease Registry). Toxicological profile for polycyclic aromatic hydrocarbons (PAHs); US Department of Health and Human Services, Public Health Service: Atlanta, GA, 1995, http://www. atsdr.cdc.gov/toxprofiles/tp69.pdf.

Ayres, E.; van der Wal, R.; Sommerkorn, M.; Bardgett, R. D. *Biol. Lett.,* 2006, 2, 286-288.

Bargagli, R.; Monaci, F.; Borghini, F.; Bravi, F.; Agnorelli, C. *Environ. Pollut.* 2002, 116, 279-287.

Bates, J. W. *J. Bryol.,* 1992, 17, 223-240.

Becker, W.; Reece, J.; Poenie, M. The World of the Cell; The Benjamin/Cumming Publishing Company: California, USA, 1995.

Belevermiş, M.; Çotuk, Y. *J. Environ. Radioactiv.,* 2010, 101, 945-951.

Berg, T.; Steinnes, E. *Environ. Pollut.,* 1997, 98, 61-71.

Boryło, A.; Nowicki, W.; Olszewski, G.; Skwarzec, B. *J. Environ. Sci. Health A Toxic/Hazard. Subst. Environ. Eng.*, 2012, 47, 1831-1842.

Brown, D. H.; Bates, J. W. Bot. *J. Linn. Soc.,* 1990, 104, 129-147.

Burger, J.; Gochfeld, M. *Environ. Monit. Assess.,* 2001, 66, 23-46.

Burton, M. A. S. *Bryologist,* 1979, 82, 594-598.

Cabrerizo, A.; Dachs, J.; Barceló, D.; Jones, K. C. *Environ. Sci. Technol.,* 2012, 46, 1396-1405.

Carballeira, A.; Fernandez, J. A.; Aboal, J. R.; Real, C.; Couto, J. A. *Atm. Environ.,* 2006, 40, 5776-5786.

Cerbunis, D.; Steinnes, E. *Atm. Environ,* 2000, 34, 4265-4271.

Choe, H.; Sackin, H.; Palmer, L. G. *J. Gen. Physiol.,* 2000, 115, 391-404.

Clymo, R. S. In Production Ecology of British Moors and Montane Grasslands; Heal, O.W.; Perkins, D.F.; Eds.; Springer-Verlag: Berlin, 1978, pp 187-223.

Ćujić, M.; Dragović, S.; Sabovljević, M.; Slavković-Beškoski, L.; Kilibarda, M.; Savović, J.; Onjia, A. *CLEAN - Soil Air Water*, 2013, DOI: 10.1002/clen.201100656.

Culicov, O. A.; Frontasyeva, M.V.; Steinnes, E.; Okina, O. S.; Santa, Z.; Todoran, R. *J. Radioan. Nucl. Chem.*, 2002, 254, 109-115.

Delfanti, R.; Papucci, C.; Benco, C. *Sci. Total Environ.* 1999, 227, 49-56.

Demidchik, V. V.; Naidun, S. N.; Yablontskaya, L. I.; Sokolik, A. I.; Yurin, V. M. *Russ. J. Plant. Physiol.*, 2001, 48, 294-299.

Dołęgowska, S.; Migaszewski, Z. M. Ecotox. *Environ. Safety* 2011, 74, 1636-1644.

Draber, S.; Hansen, U.-P. *Biophys. J.*, 1994, 64, 120-129.

Dragović, S.; Howard, B. J.; Caborn, J. A.; Barnett, C. L.; Mihailović, N. *Environ. Monit. Assess.*, 2010a, 166, 677-686.

Dragović, S.; Mandić, Lj. *Radiat. Environ. Biophys.*, 2010, 49, 625-634.

Dragović, S.; Mihailović, N. *Environ. Monit. Assess.*, 2009, 157, 383-390.

Dragović, S.; Mihailović, N.; Gajić, B. *J. Environ. Radioactiv.*, 2010b, 101, 159-164.

Dragović, S.; Nedić, O. In Radioactive Contamination Research Developments; Henshaw, N. K.; Alleyne, C. S.; Eds.; Nova Science Publishers, Inc.: New York, USA, 2010, pp 43-66.

Dragović, S.; Nedić, O.; Stanković, S.; Bačić, G. *J. Environ. Radioactiv.*, 2004, 77, 381-388.

Dragović, S.; Onjia, A.; Dragović, R.; Bačić, G. *Environ. Monit. Assess.*, 2007, 130, 245-253.

Dragović, S.; Stanković, S.; Nedić, O. *J. Serb. Chem. Soc.*, 2002, 67, 587-591.

Elstner, E. F.; Fink, R.; Höll, W.; Lengfelder, E.; Ziegler, H. *Oecologia,* 1987, 73, 553-558.

Erisman, J. W.; Brydges, T.; Bull, K.; Cowling, E.; Grennfelt, P.; Nordberg, L.; Satake, K.; Schneider, T.; Smeulders, S.; Van der Hoek, K. W.; Wisniewski, J. R.; Wisniewski, J. *Environ. Pollut.*, 1998, 102, 3-12.

Ferguson, P.; Lee, J. A. *Environ. Pollut.*, 1980, 21, 59-71.

Ferguson, P.; Lee, J. A.; Bell, J. N. B. *Environ. Pollut.*, 1978, 16, 151-162.

Fernández, J. A.; Carballeira, A. *Chemosphere,* 2002, 46, 535-542.

Fernandeź J. A.; Aboal J. R.; Couto J. A; Carballeira A. *Atm. Environ.*, 2004, 38, 4319-4329.

Foan, L.; Sablayrolles, C.; Elustondo, D.; Laceras, E.; González, L.; Hederá, A.; Simon, V.; Santamaría, J. M. *Atm. Environ.*, 2010, 44, 3207-3214.

Foan, L.; Simon, V. *J. Chromat. A,* 2012, 1256, 22-31.

Fowler, D.; Pitcairn, C. E. R.; Sutton, M. A.; Flechard, C.; Loubet, B.; Coyle, M.; Munro, R. C. *Environ. Pollut.*, 1998, 102, 343-348.

Frontasyeva, M. V.; Galinskaya, T. Y.; Krmar, M.; Matavulj, M.; Pavlov, S. S.; Povtoreyko, E. A.; Radnović, D. *J. Radioan. Nucl. Chem.*, 2004, 259, 141-147.

Gałuszka, A. *Chemosphere,* 2007, 67, 1415-1422.

Gerdola, R.; Bragazza, L.; Marchesini, R.; Medici, A.; Pedrini, P.; Benedetti, S.; Bovolenta, A.; Coppi, S. *Atm. Environ.*, 2002, 36, 4069-4075.

Giovani, C. *Sci. Total Environ.*, 1994, 157, 35-43.

Gombert, S.; Rausch de Traubenberg, C.; Losno, R.; Leblond, S.; Colin, J. L.; Cossa, D. *J. Atmos. Chem.*, 2004, 49, 479-502.

Gondzáles-Miqueo, L.; Elustondo, D.; Lasheras, E.; Santamaría, J. M. *Chemosphere,* 2010, 78, 965-971.

Goodarzi, F.; Senei, H. In Proceedings of the International Conference on Coal Science and Technology (ICCS&T); Okinawa, Japan, 2005.

Gorham, E. *Nature,* 1958, 181, 1523-1524.

Goyal, M. R. D.; Seaward, R. *New Phytol.*, 1982, 90, 73-84.

Grdović, S.; Vitorović, G.; Mitrović, B.; Andrić, V.; Petrujkić, B.; Obradović, M. *Arch. Biol. Sci.*, 2010, 62, 301-307.

Grodzińska, K.; Frontasyeva, M. V.; Szarek-Łukaszewska, G.; Klich, M.; Kucharska-Fabiś, A.; Gundorina, S. F.; Ostrovnaya, T. M. *Environ. Monit. Assess.*, 2003, 87, 255-270.

Grodzińska, K.; Szarek-Łukaszewska, G. *Environ. Pollut.*, 2001, 114, 443-451.

Hall, J. L. *J. Exp. Bot.*, 2002, 53, 1-11.

Hallingbäck, T. *Biol. Conserv.*, 1992, 59, 163-170.

Harmens, H.; Norris, D. and the participants of the moss survey, Spatial and temporal trends in heavy metal accumulation in mosses in Europe (1990-2005). Programme Coordination Centre for the ICP Vegetation; Centre for Ecology and Hydrology: Bangor, UK. 2008, ISBN 978-1-85531-239-5.

Harmens, H.; Norris, D.; Mills, G. and the participants of the moss survey. Heavy metals and nitrogen in mosses: spatial patterns in 2010/2011 and long-term temporal trends in Europe; Programme Coordination Centre for the ICP Vegetation, Centre for Ecology and Hydrology; Bangor, UK. 2013, ISBN 978-1-906698-38-6.

Harmens, H.; Norris, D. A.; Cooper, D. M.; Mills, G.; Steinnes, E.; Kubin, E.; Thöni, L.; Aboal, J. R.; Alber, R.; Carballeira, A.; Coşkun, M.; De Temmerman, L.; Frolova, M.; González-Miqueo, L.; Jeran, Z.; Leblond, S.; Liiv, S.; Maňkovská, B.; Pesch, R.; Poikolainen, J.; Rühling, Å.; Santamaria, J. M.; Simonèiè, P.; Schröder, W.; Suchara, I.; Yurukova, L.; Zechmeister, H. G. *Environ. Pollut.*, 2011, 159, 2852-2860.

Hicks, W. K.; Leith, I. D.; Woodin, S. J.; Fowler, D. *Environ. Pollut.*, 2000, 107, 367-376.

Hranisavljević-Jakovljević, M.; Miljković-Stojanović, J.; Dimitrijević, R; Mićović, V. *Arch. Carbohyd. Res.*, 1980, 80, 291-295.

Hvinden, T.; Lillegraven, A. *Nature,* 1961, 192, 1144-1146.

Knulst, J. C.; Olle Westling, H.; Brorström-Lundén, E. *Environ. Monit. Assess.*, 1995, 36, 75-91.

Krmar, M.; Radnovic, D.; Rakic, S.; Matavuly, M. *J. Environ. Radioactiv.*, 2007, 95, 53-61.

Krmar, M.; Wattanavatee, K.; Radnović, D.; Slivka, J.; Bhongsuwan, T.; Frontasyeva, M. V.; Pavlov, S. S. *J. Environ. Radioactiv.*, 2013, 117, 45-48.

Krommer, V.; Zechmeister, H. G.; Roder, I.; Scharf, S.; Hanus-Illnar, A. *Chemosphere,* 2007, 67, 1956-1966.

Leith, I. D.; van Dijk, N.; Pitcairn, C. E. R.; Wolseley, P. A.; Whitfield, C. P.; Sutton, M. A. Biomonitoring Methods for Assessing the Impacts of Nitrogen Pollution: Refinement and Testing. Report 386; Joint Nature Conservancy Committee: Peterborough, UK, 2005; http://www.jncc.gov.uk/page-3886.

Liu, X. Y; Xiao, H. Y.; Liu, C. Q.; Xiao, H. W.; Wang, Y. L. *Environ. Pollut.*, 2009, 157, 2066-2071.

Lucaciu, A.; Timofte, O.; Culicov, O.; Frontasyeva, M. V.; Oprea, C.; Cucu-Man, S.; Mocanu, R.; Steinnes, E. *J. Atmos. Chem.* 2004, 49, 533-548.

Markert, B.; Weckert, V. *Sci. Tot. Environ.*, 1989, 86, 289-94.

Markert, B. A.; Breure, A. M.; Zechmeister, H. G. In Bioindicators and Biomonitors; Markert, B. A.; Breure, A. M.; Zechmeister, H. G.; Eds.; Elsevier: Amsterdam, 2003, pp 3-39.

Markert, B.; Oehlmann, J.; Roth, M. In Environmental Biomonitoring-Exposure Assessment and Specimen Banking; Subramanian, K. S.; Iyengar, G. V.; Eds.; ACS Symposium Series 654; American Chemical Society: Washington D.C., 1997; pp 18-29.

Migaszewski, Z. M.; Gałuszka, A.; Crock, J. G.; Lamothe, P. J.; Dołęgowska, S. *Atm. Environ.*, 2009, 43, 1464-1473.

Milukaite, A. *Water, Air Soil Pollut.*, 1998, 105, 471-480.

Nedić, O.; Dragović, S. *Phytochem. Anal.*, 2006, 17, 204-207.

Nifontova, M. *Sci. Total Environ.*, 1995, 160/161, 749-752.

Nriagu, J. O.; Glooschenko, W. A. *Environ. Sci. Technol.*, 1992, 26, 85-89.

Oldfield, F.; Appleby, P. G.; Cambray, R. S.; Eakins, J. D.; Barber, K. E.; Battarbee, R. W.; Person, G. R.; Williams, J. M. *Oikos,* 1979, 33, 40-45.

Oliver, J. G. J.; Bouwman, A. F.; Van der Hock, K. W.; Berdowski, J. J. M. *Environ. Pollut*. 1998, 102, 135-148.

Orliński, R. *Chemosphere,* 2002, 48, 181-186.

Ötvös, E.; Kozák, I. O.; Fekete, J.; Sharma; V. K.; Tuba, Z. *Sci. Total Environ.,* 2004, 330, 89-99.

Ötvös, E.; Pázmándi, T.; Tuba, Z. *Sci. Tot. Environ.,* 2003, 309, 151-160.

Papastefanou, C.; Manolopoulou, M.; Sawidis, T. *J. Environ. Radioactiv.,* 1989, 9, 199-207.

Pearce, I.; Van der Val, R. *Biol. Conserv.,* 2002, 104, 83-89.

Pitcairn, C.; Fowler, D.; Leith, I.; Sheppard, L.; Tang, S.; Sutton, M.; Famulari, D. *Environ. Pollut.,* 2006, 144, 941-950.

Pitcairn, C. E. R.; Fowler, D.; Grace, J. *Environ. Pollut.,* 1995, 88, 193-205.

Poikolainen, J. Mosses, epiphytic lichens and tree bark as biomonitors for air pollutants - specifically for heavy metals in regional surveys. PhD Thesis; Faculty of Science, University of Oulu: Oulu, Finland, 2004; pp 11-21.

Popic, J. M.; Salbu, B.; Strand, T.; Skipperud, L. *J. Environ. Monit.,* 2011, 13, 1730-1738.

Popović, D.; Todorović, D.; Frontasyeva, M.; Ajtić, J.; Tasić, M., Rajšić, S. *Environ. Sci. Pollut. Res.,* 2008, 15, 509-520.

Popović, D.; Todorović, D.; Nikolić, J. *Nucl. Radiat. Radiat. Prot.,* 2009, 24, 100-103.

Puxbaum, H. In Metals and their Compounds in the Environment. Occurrence, Analysis and Biological Relevance; Merian, E.; Ed.; VCH: Weinheim, 1991, pp 176-257.

Rahman, M. M.; Voigt, V. *J. Environ. Radioact.* 2004, 71, 127-138.

Ravindra, K.; Sokhi, R.; Van Grieken, R. *Atm. Environ.,* 2008, 42, 2895-2921.

Raymond, B. A.; Bassingthwaighte, T.; Shaw, D. P. *J. Limnol.,* 2010, 69, 22-32.

Reimann, C.; Kashulina, G.; De Caritat, P.; Niskavaara, H. *Appl. Geochem.,* 2001, 16, 759-780.

Rühling, Å.; Tyler, G. *Bot. Notiser,* 1968, 122, 248-342.

Rühling, Å.; Tyler, G. *Oikos.,* 1970, 21, 92-97.

Sabovljević, M.; Vukojević, V.; Mihailović, N.; Dražić, G.; Vučinić, Ž. *Arch. Biol. Sci.,* 2005, 57, 205-212.

Sabovljević, M.; Vukojević, V.; Sabovljević, A.; Mihailović, N.; Dražić, G.; Vučinić, Ž. *Arch. Biol. Sci.,* 2007, 59, 351-361.

Sabovljević, M.; Vukojević, V.; Sabovljević, A.; Vujičić, M. *J. Ecol. Natur. Environ.* 2009, 1, 147-155.

Salemaa, M.; Mäkipaää, R.; Oksanen, J. *Environ. Pollut.,* 2008, 152, 82-91.

Shimwell, D. W.; Laurie, A. E. *Environ. Pollut.*, 1972, 3, 291-301.
Skaar, H.; Ophus, E.; Gullvag, G. M. *Nature,* 1973, 241, 215-216.
Sloof, J. E. Environmental lichenology: Biomonitoring trace-element air pollution, Delft University of Technology: The Netherlands, 1993.
Sloof, J. E.; de Bruin, M.; Wolterbeek, H. Th. In Proc. Int. Conf. Environmental Contamination, Venice; Orio A. A.; Ed.; CEP Consultants Ltd.: Edinburgh, 1988, pp 296-298.
Solga, A.; Burkhardt, J.; Zechmeister, H. G.; Frahm, J. P. *Environ. Pollut.*, 2005, 134, 465-473.
Steinnes, E. In Control and Fate of Atmospheric Trace Metals; Pacyna J. M.; Ottar B.; Eds.; Kluwer Academic Publishers: Dordrecht, 1989, pp 321-338.
Steinnes, E.; Njåstad, O. *J. Environ. Radioactiv.*, 1993, 21, 65-73.
Suchara, I.; Sucharová, J. *J. Atmos. Chem.*, 2004, 49, 503-519.
Sucharová, J.; Suchara, I. *Chemosphere,* 2004, 57, 1389-1398.
Svensson, G. K.; Linden, K. *Health Physiol.*, 1965, 11, 1033-1042.
Szarek, G.; Chrzanowska, E. *Acta Soc. Bot. Polon.*, 1991, 60, 101-110.
Thomas, W. *Ecotox. Environ. Safety,* 1986, 11, 339-346.
Tyler, G. Proceedings of Second International Clean Air Congress; Washington: USA, 1970.
Uğur, A.; Özden, B.; Saç, M. M.; Yener, G. *Atm. Environ.*, 2003, 37, 2237-2245.
Uğur, A.; Özden, B.; Saç, M. M.; Yener, G.; Altınbaş, Ü.; Kurucu, Y.; Bolca, M. *J. Radioan. Nucl. Chem.* 2004, 259, 87-92.
USEPA (United States Environmental Protection Agency). List of Chemicals Evaluated for Carcinogenic Potential; Office of Pesticide Programs: Washington, DC, USA, 1997.
Viet, H. N.; Frontasyeva, M. V.; Trinh Thi, T. M.; Gilbert, D.; Bernard, N. *Environ. Sci. Pollut. Res.* 2010, 17, 1045-1052.
Vukojević, V.; Sabovljević, M.; Sabovljević, A.; Mihailović, N.; Dražić, G.; Vučinić, Ž. *Arch. Biol. Sci.* 2006, 58, 95-104.
Vukojević, V.; Sabovljević, M.; Sabovljević, A.; Mihailović, N.; Dražić, G.; Vučinić, Ž. *Arch. Biol. Sci.* 2009, 61, 835-845.
Wagner, G. Einsatzstrategien und Meßnetze für die Bioindikation im Umweltmonitoring; Ecoinforma: Bayreuth, 1992.
White, P. J.; Broadley, M. R. *New Phytol.* 2000, 147, 241-256.
Xiao, H. Y.; Zhu, R. G.; Lin, B. N.; Liu, C. Q. *Sci. Total Environ..* 2011, 409, 2127-2132.

Zechmeister, H. G.; Grodzinska, K.; Szarek-Lukaszewska, G. In Bioindicators and Biomonitors; Markert, B. A.; Breure, A. M.; Zechmeister, H. G.; Eds.; Elsevier Science Ltd: Amsterdam, Netherlands, 2003, pp 329-375.

Zechmeister, H. G.; Richter, A.; Smidt, S.; Hohenwallner, D.; Roder, I.; Maringer, S.; Wanek, W. *Environ. Sci. Technol.* 2008, 42, 8661-8667.

Zhu, Y. -G.; Smolders, E. *J. Exp. Bot.* 2000, 51, 1635-1645.

In: Moss
Editor: Jorma Mohamed

ISBN: 978-1-63117-396-7

Chapter 7

IMPACT OF FOREST DECLINE ON THE FINE-SCALE VERTICAL DISTRIBUTION OF EPIPHYTIC BRYOPHYTES

Yoshitaka Oishi*
Faculty of Agriculture, Shinshu University, 8304 Minami-minowa, Kami-ina, Nagano, Japan

ABSTRACT

The vertical distribution of epiphytic bryophytes on tree trunks may serve as an indicator of diversity, and requires delineation. This study assessed changes in the vertical distribution of epiphytic bryophytes to determine how forest decline affects the diversity of these non-vascular plants. Epiphytic bryophytes growing on the trunks of *Pinus jezoensis* var. *hondoensis* trees were examined along four 50-cm-high zones from ground level to a trunk height of 2 m in six forest patches at different states of forest decline (no decline, slight decline, severe decline, and serious decline). Overall, epiphyte species richness and trunk coverage decreased with the process of forest decline. Using two-way indicator species analysis and detrended correspondence analysis, epiphytic bryophyte flora diversity appeared to shift from higher trunk zones in healthy forests to lower trunk zones in heavily degraded forests. These changes might be explained by an increase in drought stress and light intensity caused by forest decline, which might have a greater effect on

* Corresponding author: Email: oishiy@shinshu-u.ac.jp.

the composition of epiphytic bryophytes inhabiting the lower zones of trunks. Therefore, the epiphytic bryophytes occupying the lower trunk zones in forest patches where forest decline continues require careful monitoring and conservation. As the fine-scale distribution of epiphytic bryophytes may be easily surveyed without any special equipment or machines, these plants could be used to regularly monitor and evaluate the impact of forest decline on epiphytic bryophytes.

1. INTRODUCTION

Forest decline has a serious negative impact on the biodiversity contained within forest ecosystems (Mueller-Dombois 1988). The decrease in tree coverage increases the exposure of forests to strong winds and raises the light intensity that penetrates the forest floor, both of which cause a decrease in the humidity of forests (Ando 2009). These changes are expected to have a major effect on epiphytic bryophytes, particularly species that grow on the lower sections of tree trunks, because they preferentially grow in humid habitats and are vulnerable to desiccation (Franks & Bergstrom 2000; Fuertes et al. 1996). Furthermore, the rounded shape of tree trunks might also enhance the effects of drought stress (Baldwin & Bradfield 2007; Hylander et al. 2005). Epiphytes generally represent a small proportion of the total forest biomass; however, they are critical for nutrient cycling, because they efficiently intercept and retain nutrients from the atmosphere (Nadkarni 1984). Epiphytes also influence forest hydrology, because they store large quantities of water; thus increasing canopy water-storage levels (Pypker et al. 2006).

Various environmental factors influence the diversity of epiphytic bryophytes. Such factors include tree density (Ojala et al. 2000), tree age (Fritz 2009), height above the ground (Fritz 2009; Fuertes et al. 1996; Hosokawa & Odani 1957; Sillett & Rambo 2000), host tree diameter at breast height [DBH] (Hazell et al. 1998; Ojala et al. 2000), bark features (Hale 1952; Kenkel & Bradfield 1981), bark pH (Mežaka & Znotiņa 2006), air purity (Giordano et al. 2004), past landscape structure (Snäll et al. 2004), past deforestation (Holz et al. 2005), and demographic processes (Pereira Alvarenga & Pôrto 2007; Pharo & Zartman 2007). Of these parameters, the height above ground level might have a major influence on bryophyte diversity, because bryophyte distribution is likely to be influenced by vertical changes in environmental conditions that are important for growth, such as light, moisture, and bark features (Fritz 2009; Lara & Mazimpaka 1998; McCune et al. 1997; Sillett & Rambo 2000; Sporn et al. 2010). In general, the humidity on tree trunks decreases with increasing trunk height above ground level, because of increased exposure to desiccating winds and sunlight. Consequently, xeric species preferentially grow on the upper part of tree trunks, whereas hygrophilous species preferentially grow on the lower part (Ellyson & Sillett 2003; Franks & Bergstrom 2000; Fritz 2009; Iwatsuki 1960; Lara & Mazimpaka 1998; Lyons et al. 2000). This high humidity, in combination with the physical characteristics of tree bark, results in higher bryophyte diversity occurring near the tree base (Mežaka & Znotiņa 2006). Hosokawa and Odani (1957) showed that the vertical range of epiphytes is strongly associated with the daily compensation period, which might be an important factor influencing the vertical distribution of epiphytes.

Changes in the vertical distribution of epiphytes have been reported for several different types of forests (Franks & Bergstrom 2000; Fritz 2009; Hale 1952; Holz et al 2005; McCune et al. 1997; Sillett & Rambo 2000; Sporn et al 2010). However, only a few studies have documented how differences or changes in environmental conditions affect the vertical distribution of epiphytes (Acebey et al. 2003). Acebey et al. (2003) compared epiphyte diversity in a primary rain forest versus young fallows (i.e., the early stages of forest regeneration following transient agricultural use). The authors found that the highest species richness in the primary forest was in the canopy, whereas that in the fallows was in the understory. This difference was explained by the warmer air temperature and lower humidity in the understory of the fallows compared to the primary forest.

Mount Odaigahara is an important biodiversity forest reserve in Japan; however, this reserve has been subject to ongoing forest decline since the 1980s, with trees dying following bark stripping activity by deer (Ando & Goda 2009; Yokoyama et al. 2001). The vegetation in this area is characterized by high epiphytic bryophyte diversity, because of high humidity conditions caused by high precipitation, which exceeds 4500 mm annually (Doei 1988).

Previous studies (Oishi 2011, 2012, 2013) indicated that the diversity of epiphytic bryophytes on the lower part of tree trunks might be negatively influenced by forest decline. Nevertheless, to date, the impact of forest decline on the vertical distribution of these epiphytic bryophytes has not been evaluated; therefore, it is important to quantify how forest decline affects their vertical distribution, and hence their diversity. In this study, we specifically compared the vertical distribution of epiphytic bryophytes on lower part of tree trunks in adjacent forest patches at different stages of forest decline. The results of the current study are expected to contribute toward improving our understanding about the impact of forest decline on bryophyte diversity. Ultimately, this study might also contribute information to the conservation of forest ecosystems through elucidating the ecological functions of bryophytes.

2. Materials and Methods

2.1. Study Sites and Study Plots

Mount Odaigahara (Longitude 34°N, Latitude 136°E) is situated within the Yoshino Kumano National Park, which is located in the southeastern area of Nara Prefecture in the Kii Peninsula, Japan (Figure 1, revised from Oishi 2011). This mountain is an example of an uplifted peneplain. The height of this area ranges from 1400 m in the west to 1600 m in the east, with gentle slopes. Different vegetation grows on the eastern and western parts of the mountain. Specifically, the eastern part is mainly covered by *P. jezoensis* forests, whereas the western part is mainly covered by *Fagus crenata* Blume and *Abies homolepis* Siebold et Zucc. The extent of forest decline is more serious on the eastern part of the mountain, because *P. jezoensis* trees are prone to heavy damage caused by deer debarking activity (Ando et al. 2003). The climate in this region is characterized by cool temperatures and heavy precipitation (Nara Local Meteorological Observatory 1997).

On Mt. Odaigahara, tree trunks of approximately 32,500 trees were wrapped with wire mesh to prevent bark stripping by deer (Ministry of the Environment, Kinki Regional Environment Office 2009); however, previous research has demonstrated that wire mesh protection has a significant negative effect on bryophyte diversity through metal pollution (Oishi 2011, 2012). Therefore, trees with wire mesh protection were excluded from the current study to examine the influence of forest decline on the vertical distribution of epiphytic bryophytes. Consequently, six 20 m × 20 m quadrats (Plots 1, 2, 3, 4, 5, and 6), located in forest patches at different stages of forest decline, were selected as the study plots. The locations of the plots and associated environmental conditions are presented in Figure 1 and Table 1, respectively. Plots 1, 2, and 3 were categorized as no declined, slightly declined, and severely declined forest, respectively; whereas plots 4, 5, 6 were classified as seriously declined forest. Oishi (2011, 2012, 2013) has previously reported on the relationships between bryophyte diversity and environmental conditions in these plots.

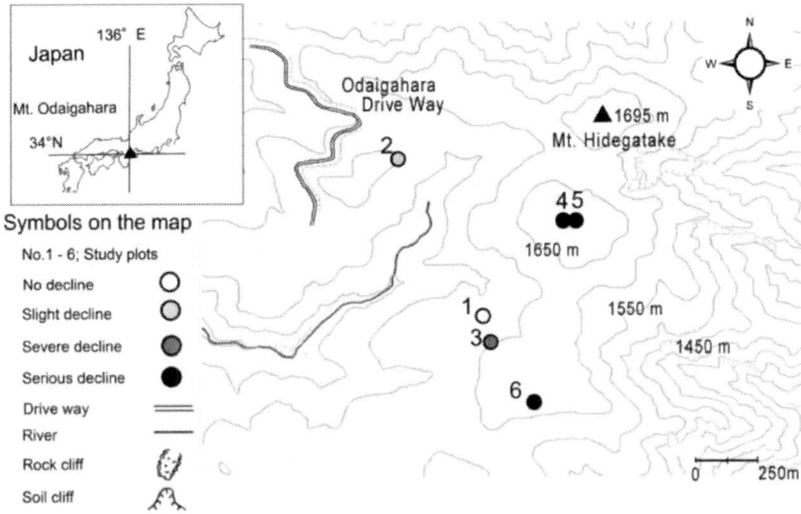

Figure 1. Location of Mount Odaigahara and the six study plots. Mount Odaigahara is located in the Kii Peninsula in the Kinki district of central Japan. Revised from Oishi (2011).

Table 1. Overview of the environmental conditions in each study plot. Four forest decline categories were delineated: no decline, with >75% tree canopy cover; slight decline, with 50–75% tree canopy cover; severe decline, with 25–50% tree canopy cover; and serious decline, with <25% tree canopy cover

Plot	Altitude (m)	Forest decline
Plot 1	1572	no decline
Plot 2	1597	slight
Plot 3	1597	severe
Plot 4	1676	serious
Plot 5	1672	serious
Plot 6	1621	serious

Figure 2. The four forest decline categories. A, No decline; B, Slight decline; C, Severe decline; D, Serious decline.

2.2. Epiphyte Sampling

The epiphyte flora on the trunks of *P. jezoensis* var. *hondoensis* trees in the plots were subject to detailed surveys from October to November 2008 (Oishi 2011). All epiphyte species growing on tree trunks were recorded and categorized into four zones above ground level: 0–50 cm (zone A), 50–100 cm (zone B), 100–150 cm (zone C), and 150–200 cm (zone D). Epiphytic bryophytes on trunks covered with soil were not recorded. The coverage of each species was recorded using a 5 cm × 5 cm quadrat and classified according to the Braun-Blanquet cover-abundance scale (Mueller-Dombois & Ellenberg 1974). In addition, the percentage of total bryophyte coverage relative to the basal area was measured. Epiphyte species that could not be identified in the field were collected and transferred to the laboratory for identification under a microscope. The nomenclature followed that of Iwatuski (2001).

2.3. Measurements of Environmental Conditions

The degree of forest decline was classified into the following four categories (Figure 2): (1) no decline, with >75% tree canopy cover; (2) slight decline, with 50–75% tree canopy cover; (3) severe decline, with 25–50% tree canopy cover; and (4) serious decline, with <25% tree canopy cover). The diameter at breast height (DBH) of all trees was measured. In the following analysis, trees without wire mesh protection were used to examine the relationship between forest decline and the vertical distribution of epiphytic bryophytes.

2.4. Changes in Epiphyte Diversity in Relation to Height Zone

Changes in bryophyte species richness and coverage on the four tree trunk height zones were compared in a single forest that contained patches at different stages of decline. In this comparison, the dominance of bryophyte life-forms was used to characterize differences in species composition. Life-forms are usually interpreted as plants that are adapted to certain moisture and light intensity conditions, which represent important factors for bryophyte growth (Bates 1998). Consequently, the changes in life-form composition are potentially useful for evaluating how these environmental factors influence epiphyte diversity. The life-forms used here were selected using the classification of Bates (1998).

Seven life-forms were observed in the study plots; namely, short turfs, tall turfs, small cushions, smooth mats, thread-like forms, wefts, and fans. Open life forms, such as fans and wefts, usually occupy shaded and moist environments (Bates 1998). Tall turfs also tend to grow in moist environments and on hard substrata (Bates 1998). Therefore, these three life forms may be considered as the hygrophilous life forms, among the seven observed life forms. In contrast, small cushions are highly desiccation tolerant (Bates 1998).

The number of trees sampled in the four forest categories differed greatly (from just three in the seriously declined forest to 22 in the severely declined forest); therefore, the average species richness per tree in each forest category was used for the comparison of bryophyte diversity.

2.5. Species Composition

To determine how bryophyte species composition differs with different levels of forest decline, the study plots were first classified into groups with similar species composition based on two-way indicator species analysis (TWINSPAN). Following TWINSPAN analysis, detrended correspondence analysis (DCA) was conducted to examine differences in species composition. Bryophyte species that were only recorded up to two times in a given plot were excluded from the analysis. TWINSPAN analysis was based on the mean coverage of each bryophyte species (Braun-Blanquet cover-abundance scale) using seven pseudospecies levels: 0, 0.1, 1, 2, 3, 4, and 5. In the DCA ordination, rare species were down-weighted, otherwise the analysis followed the default options.

After DCA analysis, correlation analyses (Pearson product-moment correlation coefficient) were carried out between environmental and ecological variables, with the DCA axes being used to identify the vegetation gradient and associated environmental changes. The environmental variables used in this analysis were the level of forest decline, mean DBH of host trees, and height zone category. The ecological variables used in this analysis were bryophyte species richness, total bryophyte coverage, and the dominance of each bryophyte life-form in each of the four height zones. The dominance of life-forms was calculated as the ratio of bryophyte species richness of each life-form to total bryophyte species richness for each height zone. The TWINSPAN and DCA analyses were performed using PC-ORD version 5.10 (McCune & Mefford 2006).

The average species richness per tree in each plot was used for these analyses because of the large difference in tree numbers among plots (Range per plot: 1–22 trees) risked generating biased results. Therefore, 21 samples (4 height classes × 6 plots = 24 combined samples, then minus three that had no bryophytes) were used in the TWINSPAN and DCA analyses.

3. RESULTS

3.1. Changes in Epiphyte Diversity

Fifty-eight species were identified in the epiphyte flora survey (including 25 moss and 33 liverwort species) on the trees in the six study plots. The frequently occurring species were *Dicranum leiodontum* Card., *Dicranum viride* (Sull. & Lesq.) Lindb. var. *hakkodense* (Card.) Takaki, *Heterophyllium affine* (Hook.) M.Fleisch,.*Hypnum tristo-viride* (Broth.) Paris, *Herbertus aduncus* (Dicks.) Gray, *Bazzania denudata* (Torr. ex Lindenb.) Trevis., *Frullania tamarisci* (L.) Dumort. subsp. *obscura* (Verd.) S.Hatt., and *Nipponolejeunea pilifera* (Steph.) S. Hatt. (Figure 3). Bryophyte species were not recorded in tree trunk height zones A and B in plot 5, or zone D in plot 6.

Changes in mean species richness and cover per tree along the tree trunk height zones, and according to forest decline, are presented in Figure 4. Overall, mean species richness and cover were highest on tree trunk height zone A, and declined with each increment in height zone. Mean species richness and coverage declined with increasing level of forest decline. This trend was prominent in zone A.

Figure 3. Major bryophyte species in the study plot: A, Dicrunum viride var. hakkodense; B, Heterophyllium affine; C, Hypnum tristo-viride; D, Nipponolejeunea pilifera.

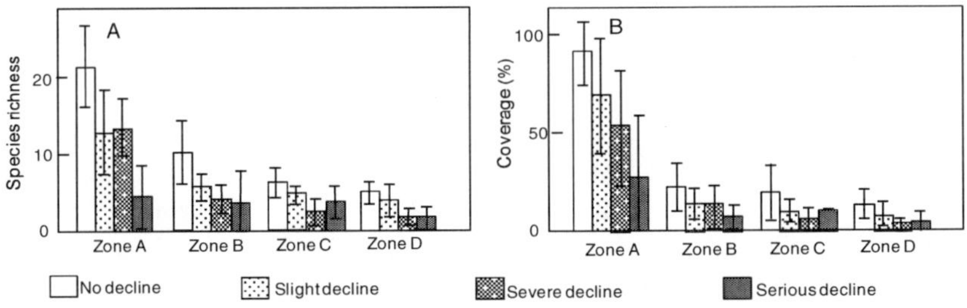

Figure 4. Comparison of the species richness (A) and coverage (B) of epiphytes for each height zone (Zone A–D) on Pinus jezoensis var. hondoensis trunks in the forest plots with different levels of forest decline. Bars represent one standard error. Height zones above ground level: A, 0–50 cm; B, 50–100 cm; C, 100–150 cm; D, 150–200 cm. The classes of forest decline are defined in Table 1.

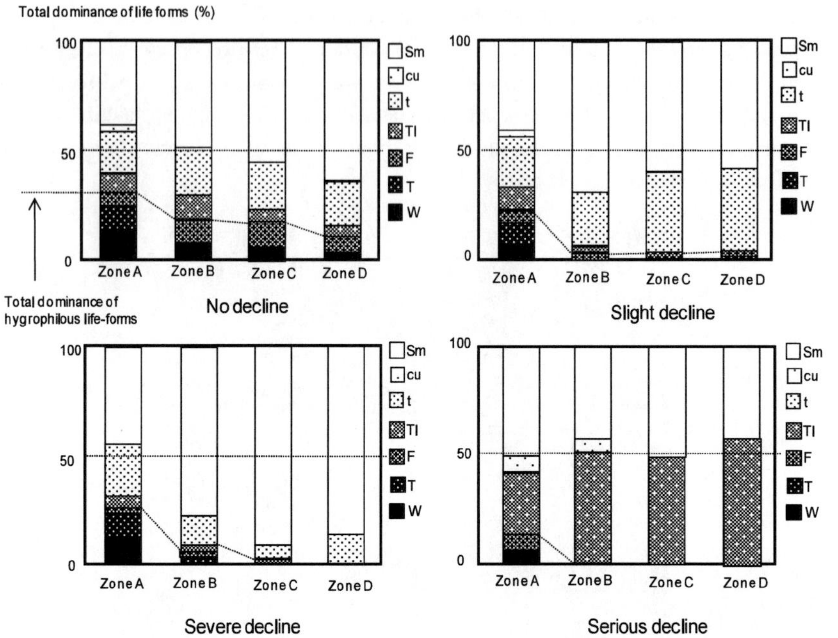

Figure 5. Comparison of the relative dominance of bryophyte life-forms for the four tree trunk height zones in the forest plots with different levels of forest decline. The tree trunk height zones and the category of forest decline are defined in Figure 4 and Table 1, respectively. The bryophyte life-form abbreviations are as follows; t = short turfs; T = tall turfs; cu = small cushions; Sm = smooth mats; Tl = thread-like forms; W = wefts, and F = fans.

Differences in life-form composition for both the tree trunk height zone and the level of forest decline are presented in Figure 5. The presence of tall turfs and wefts decreased in the upper tree trunk height zones. The total dominance of hygrophilous life-forms (wefts, tall turfs, and fans) was the greatest at all tree trunk height zones in forests with no decline, and was the lowest in forests with serious decline.

3.2. Species Composition

Using TWINSPAN, the six study plots were classified into three groups (Groups A–C; Figure 6). Group A mainly contained plots with tree trunk height zone A, except for the four plots in the forest area with serious decline.

Group B contained plots with trunk height zones B–D in forests with no decline, slight decline, and severe decline. Group C mostly contained the forest plots in the serious decline category. As shown in Figure 6, both species composition and species richness were affected by the extent of forest decline and height zone. Higher species richness was observed in forest categories with no or slight decline and lower height zones.

DCA was used to identify the relationship between species composition and environmental parameters (Figure 7). DCA ordination had a total inertia of 1.5. The first axis had the highest eigenvalue (0.50), whereas the eigenvalues of axes 2 and 3 were 0.31 and 0.04, respectively. The low eigenvalue of axis 3 indicated that it was of minor importance for explaining the species composition. The gradient lengths of axes 1 and 2 were 3.0 and 2.5, respectively.

The correlation analyses between the ordination scores and environmental variables are shown in Table 2. DCA axis 1 was strongly and positively correlated with the process of forest decline (r = 0.71), whereas DCA axis 2 was strongly correlated with height zone (r = 0.49). The correlation between the ordination scores and ecological variables showed that DCA axis 1 was significantly and strongly negatively correlated with species richness (r = –0.64) and coverage (r = –0.56). DCA axis 2 was also significantly negatively correlated with coverage (r = –0.50). With respect to the dominance of life forms, DCA axis 1 showed a significant positive correlation with the dominance of short turfs (r = 0.68), and a significant negative correlation with the dominance of tall turfs and wefts (r = –0.69 and – 0.52, respectively). DCA axis 2 showed a significant positive correlation with the dominance of smooth mats (r = 0.48), and a significant negative correlation with the dominance of wefts (r = –0.53).

4. Discussion

The results of this study indicate that forest decline has a negative effect on epiphytic bryophyte diversity, causing the vertical distribution of life-forms to change. These impacts are examined in this section, in relation to changes in species richness and composition, along with the dominance of certain life-forms (groups of species).

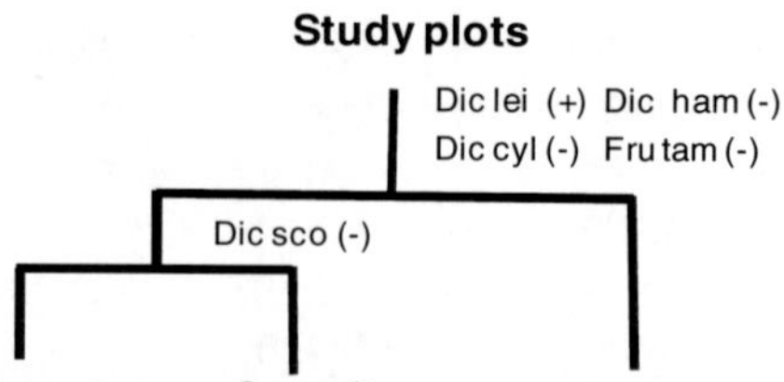

	Group A				Group B						Group C										
Forest decline	0	0	1	2	0	0	1	1	2	2	3	3	3	3	3	3	3	2	1	3	3
Height zones	A	B	A	A	C	D	B	C	B	D	B	C	A	B	C	D	D	C	D	A	C
Pog alp	2	-	1	1	-	-	-	-	-	-	-	-	-	-	-	-	-	-	-	-	-
Dic sco	2	1	1	2	-	-	-	-	-	-	-	-	-	-	-	-	-	-	-	-	-
Sca hir	1	-	1	1	-	-	-	-	-	-	-	-	-	-	-	-	-	-	-	-	-
Pla sat	1	1	-	1	-	-	-	-	-	-	-	-	-	-	-	-	-	-	-	-	-
Bro fau	1	-	1	1	-	-	-	-	-	-	-	-	-	-	-	-	-	-	-	-	-
Bro hen	1	-	1	1	-	-	-	-	-	-	-	-	-	-	-	-	-	-	-	-	-
Pse poh	1	-	1	1	-	-	-	-	-	-	-	-	-	-	-	-	-	-	-	-	-
Lep rep	2	-	2	1	-	-	-	-	-	-	-	-	-	-	-	-	-	-	-	-	-
Lop lon	1	1	1	1	-	-	-	-	-	-	-	-	-	-	-	-	-	-	-	-	-
Myl ver	1	1	2	-	-	-	-	-	-	-	-	-	-	-	-	-	-	-	-	-	-
Baz yos	2	1	2	2	-	-	-	-	-	-	-	-	-	-	-	-	-	-	-	-	-
Dic den	1	-	1	1	1	-	-	-	-	-	-	-	-	-	-	-	-	-	-	-	-
Sca amp	3	2	2	2	1	-	1	-	-	-	-	-	-	-	-	-	-	-	-	-	-
Ble tri	1	1	1	1	1	-	-	-	-	-	-	-	-	-	-	-	-	-	-	-	-
Dic cyl	2	2	2	1	1	1	1	-	-	1	-	-	-	-	-	-	-	-	-	-	-
Ana mic	1	1	1	1	1	-	1	-	-	-	-	-	-	-	-	-	-	-	-	-	-
Rad bru	1	1	-	-	1	-	-	-	-	-	-	-	-	-	-	-	-	-	-	-	-
Pla sem	2	2	1	1	2	2	1	-	-	1	-	-	-	-	-	-	-	1	-	-	-
Odo den	2	1	1	1	1	-	1	1	1	1	-	-	-	-	-	-	-	1	-	-	-
Fru tam	1	2	2	1	2	2	2	2	2	1	-	-	-	-	-	-	-	1	2	-	-
Het aff	4	2	3	4	1	1	1	-	2	-	2	2	-	-	-	-	-	-	-	5	-
Lep vit	2	1	2	2	-	-	-	-	-	-	-	-	-	-	-	-	-	-	-	2	-
Hyp fuj	2	-	2	2	-	-	-	-	-	-	-	-	-	-	-	-	-	-	-	3	-
Dic ham	2	2	2	2	1	1	1	1	2	2	-	-	-	-	-	-	-	1	-	3	-
Her adu	1	2	2	1	2	1	1	-	1	1	2	-	-	-	-	-	-	1	-	-	-
Pla gio c	1	2	1	1	2	1	1	-	1	-	-	-	2	-	-	-	-	1	-	-	-
Der ang	1	2	1	-	1	1	1	-	-	-	-	-	2	-	-	-	-	-	-	-	-
Baz bid	1	-	1	-	-	1	1	-	-	-	-	-	-	-	-	-	-	-	1	-	-
Lej uli	1	1	-	-	1	1	-	-	-	-	2	-	-	-	-	-	-	-	-	-	-
Hyp tri	4	2	3	4	2	2	2	2	2	2	3	2	2	2	2	2	-	2	1	3	-
Baz den	2	2	2	2	1	1	1	1	2	-	2	-	2	-	-	-	-	1	-	-	-
Dic lei	1	1	1	2	1	1	1	1	1	2	2	3	3	2	2	2	2	2	1	3	3
Dic vir	1	1	1	1	1	1	1	1	1	1	2	2	2	2	2	-	2	2	1	-	-
Nip pli	1	1	1	1	1	1	1	1	2	1	2	2	2	-	-	-	2	2	1	2	2
Nip sub	1	-	-	-	-	-	1	-	-	1	-	2	-	-	-	-	-	1	-	-	-

Figure 6. Classification of epiphyte communities based on TWINSPAN Indicator species for each division are presented in this figure. The categories of forest decline are as follows: 0 = no decline, 1 = slight decline, 2 = severe decline, 3 = serious decline. Tree trunk height zone categories are defined in Figure 4. The "+" symbol indicates that the indicator species was present in a group, whereas "-" indicates that the species was absent. Abbreviated species names are defined in the Appendix.

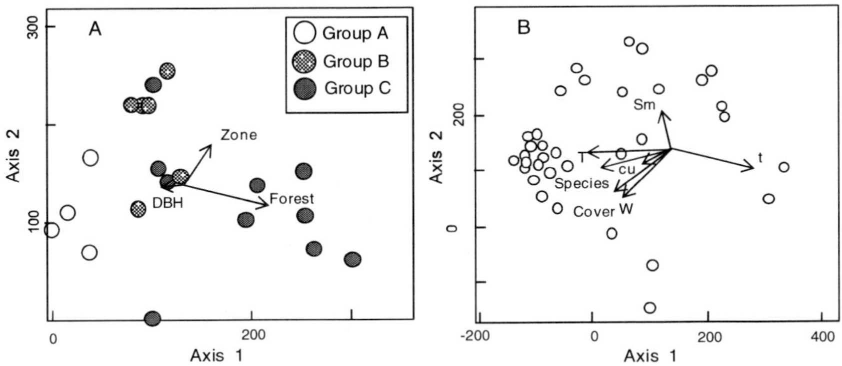

Figure 7. DCA ordination diagrams. A. Scores of all six plots along DCA axes 1 and 2; B. DCA scores of all species with more than two occurrences along DCA axes 1 and 2. The arrow shows the direction of the gradient of the environmental (A) and ecological variables (B), whereas the length of the arrow is proportional to the correlation between the variable and the ordination. The groups of plots reflect those presented in Figure 6, whereas the abbreviations of the environmental and ecological variables are presented in Figure 5 and Table 2.

4.1. Changes in Epiphytic Bryophyte Diversity along Height Zones

Comparison of the changes in species richness and coverage of epiphytes along tree trunks indicated that these values decreased with increasing tree trunk height zone. The dominance of drought-sensitive life-forms (tall turfs, fans, and wefts) decreased in the upper tree trunk height zones, which was probably caused by drought stress influencing the distance of different life-forms from the ground. Although a decrease in epiphyte diversity in upper tree trunk height zones because of drought stress has been reported in several previous studies, most studies (Fritz 2009; Hale 1952; McCune et al. 1997; Sillett & Rambo 2000; Sporn et al. 2010) discussed changes in large-scale vertical distribution (e.g., height zones: 0–2 m, 5–10 m, and 15–20 m). In contrast, only a few studies (e.g., Franks & Bergstrom 2000) examined changes in fine-scale distribution (e.g., height zones: 0–0.5 m, 0.5–1.0 m, 1.0–1.5 m, and 1.5–2.0 m). Interestingly, the results of the previous fine-scale study differed to the findings of the current study. Franks & Bergstrom (2000) reported no significant differences in mean species richness on the lower part (<2 m) of tree trunks in the microphyll fern forests of Queensland, Australia.

The different findings of the current study might be explained by the greater variation in drought stress among the six study plots (and four categories), which might be related to differences in climatic conditions (e.g., temperature and precipitation) and local environmental factors (e.g., tree density, canopy cover, and tree species) between Japan and Australia.

Table 2. Pearson's correlation between the DCA axes and environmental and ecological variables. Abbreviations: Forest = category of forest decline (see Table 1); DBH = diameter at breast height of host trees; Zone = height zones; Species = species richness of epiphytic bryophytes; Cover = total coverage of epiphytic bryophytes. The bryophyte life-form abbreviations are defined in Figure 5

	Axis 1	Axis 2
Forest	0.71**	–0.35
DBH	–0.37	–0.04
Zone	0.40	0.49*
Species	–0.64**	–0.33
Cover	–0.56*	–0.50*
Sm	–0.19	0.48*
cu	–0.40	–0.30
t	0.68**	–0.34
Tl	0.38	–0.05
F	–0.23	0.18
T	–0.69**	–0.12
W	–0.52*	–0.53*

* = p <0.05, ** = p <0.01.

4.2. Impact of Forest Decline on Species Richness

The results of the current study showed that forest decline has a negative effect on both the species richness and coverage of epiphytes at all height zones. This decrease was particularly noticeable for epiphytes in zone A. Serious damage to epiphytes in this zone might be explained by their vulnerability to severe drought stress. The decrease in tree coverage concomitant with forest decline allows more sunlight to reach the forest floor, which, in turn, causes a decline in humidity (Ando 2009). As a result, epiphytic bryophytes in the lower height zones are seriously impacted, because of their preference for growing in humid habitats (Franks & Bergstrom 2000; Fuertes et al. 1996).

4.3. Impact of Forest Decline on Species Composition

The significant strong correlations between DCA axes 1 and 2 scores and environmental conditions indicate that changes in species composition are primarily explained by the extent of forest decline and by the distance of bryophytes from the ground. Both forest decline and distance from the ground influence epiphyte diversity because of changes in moisture availability. This tendency was also shown in the classification of study plots by TWINSPAN (Figure 6).

In the DCA cross plot, the significant decrease in the dominance of hygrophilous life-forms (wefts) along DCA axes 1 and 2 indicates an increase in drought stress; however, the pattern of change in the dominance of other life-forms was not identical. This discrepancy might be related to the rapid increase in light intensity and strong drought stress caused by forest decline. These environmental changes might also cause a severe decrease in sciophilous species, such as tall turfs (e.g., *Dicranum scoparium* Hedw.; Iwatsuki 2001), resulting in an increase in small turf species, which are more tolerant to desiccation and light (Bates 1998). In contrast, because of the slight differences in distance from the ground among the four height zones, the change in light intensity and humidity along the tree trunk height zones might be smaller compared to the changes caused by forest decline.

Of note, the difference in species composition between the lower and upper height zones was obvious in forests with no decline. In contrast, this difference was not clear in the seriously declined forest category, as shown by the TWINSPAN results. The minor dominance of hygrophilous life-forms at lower tree trunk height zones in forest patches with serious decline, might indicate that epiphytic bryophyte flora shift toward the lower zones of tree trunks, because severe drought stress is equally likely to occur at lower and higher tree trunk zones. In a previous study, Acebey (2003) suggested that the shift from higher tree trunk height zones (10–15 m above the ground) to lower zones (0–5 m above the ground) occurs because of drought stress. The current study demonstrated a shift in species distribution, even on sections of tree trunk below 2 m height, because of forest decline causing a major decrease in humidity.

Conclusion

The current study demonstrates that forest decline alters the vertical distribution of epiphytic bryophytes, even at fine-scales, because of major changes in humidity and light intensity. In addition, this study shows that epiphytic bryophytes occupying the lowest tree trunk height zones (<50 cm) are particularly sensitive to forest decline. Therefore, to maintain epiphyte diversity in the face of the negative impact of forest decline, it is necessary to monitor epiphyte life-forms occupying the lower tree trunk height zones.

This study shows that the fine-scale distribution of bryophytes could be surveyed without the use of any special equipment or machinery (e.g., tower cranes); hence, these species could be used to regularly monitor and evaluate the impact of forest decline on epiphytic bryophytes. Specifically, bryophyte life-form composition might serve as a useful indicator for monitoring bryophyte diversity, because these life-forms clearly reflect the changing trends in epiphyte diversity, as well as being easy to identify in the field.

Acknowledgments

The author thanks Professor Yukihiro Morimoto for providing critical comments and suggestions for the improvement of this paper, Dr. Kosaku Yamada for assisting with bryophyte identification, and Kaori Oishi for assisting with the bryophyte survey and photos. This research was supported by Grant-in-Aid for Scientific Research (A) Grant No. 18201008 from the Japan Society for the Promotion of Science.

Appendix

Abbreviation	Species name	Abbreviation	Species name
Ana mic	*Anastrophyllum michauxii* (F.Weber) H.Buch	Hyp fuj	*Hypnum fujiyamae* (Broth.) Paris
Baz bid	*Bazzania bidentula* (Steph.) Steph.	Hyp tri	*Hypnum tristo-viride* (Broth.) Paris
Baz den	*Bazzania denudata* (Torr. ex Lindenb.) Trevis.	Lej uli	*Lejeunea ulicina* (Tayl.) Gottsche, Lindenb. & Nees

Abbreviation	Species name	Abbreviation	Species name
Baz yos	*Bazzania yoshinagana* (Steph.) S.Hatt.	Lep rep	*Lepidozia reptans* (L.) Dumort.
Ble tri	*Blepharostoma trichophyllum* (L.) Dumort.	Lep vit	*Lepidozia vitrea* Steph.
Bro fau	*Brotherella fauriei* (Card.) Broth.	Lop lon	*Lophozia longiflora* (Nees) Schiffn.
Bro hen	*Brotherella henonii* (Duby) M.Fleisch.	Myl ver	*Mylia verrucosa* Lindb.
Der ang	*Drepanolejeunea angustifolia* (Mitt.) Grolle	Nip pli	*Nipponolejeunea pilifera* (Steph.) S.Hatt.
Dic cyl	*Dicranoloma cylindrothecium* (Mitt.) Sakurai	Nip sub	*Nipponolejeunea subalpina* (Horik.) S.Hatt.
Dic den	*Dicranodontium denudatum* (Brid.) E.Britt. ex R.S.Williams	Odo den	*Odontoschisma denudatum* (Mart.) Dumort.
Dic ham	*Dicranum hamulosum* Mitt.	Pla sat	*Plagiochila satoi* S. Hatt.
Dic lei	*Dicranum leiodontum* Card.	Pla sem	*Plagiochila semidecurrens* (Lehm. & Lindenb.) Lindenb.
Dic sco	*Dicranum scoparium* Hedw.	Pla cor	*Plagiochila corniculata* (Dumort.) Dumort.
Dic vir	*Dicranum viride* (Sull. & Lesq.) Lindb. var. *hakkodense* (Card.) Takaki	Pog alp	*Pogonatum alpinum* (Hedw.) Röhl.
Fau ten	*Fauriella tenuis* (Mitt.) Card.	Pse poh	*Pseudotaxiphyllum pohliaecarpum* (Sull. & Lesq.) Z.Iwats.
Fru tam	*Frullania tamarisci* (L.) Dumort. subsp. *obscura* (Verd.) S.Hatt.	Rad bru	*Radula brunnea* Steph.
Her adu	*Herbertus aduncus* (Dicks.) Gray	Sca amp	*Scapania ampliata* Steph.
Het aff	*Heterophyllium affine* (Hook.) M.Fleisch.	Sca hir	*Scapania hirosakiensis* Steph. ex Mÿll. Frib.

References

Acebey, C., Gradstein, S. R. & Krömer, T. (2003). Species richness and habitat diversification of corticolous bryophytes in submontane rainforest and fallows of Bolivia. *Journal of Tropical Ecology*, *19*, 9–18.

Ando, M. (2009). The expansion of the grassland of *Sasa nipponica* Makino et Shibata. In: E. Shibata & T. Hino (Eds.), *Ecology of Sika deer and forest ecosystem of Mt. Ohdaigahara* (128–133). Hadano, Kanagawa: Tokai University Press.

Ando, M. & Goda, R. (2009). Sika deer (*Cervus nippon* Temminck) of Mount Ohdaigahara. In: E. Shibata & T. Hino (Eds.), *Ecology of Sika deer and forest ecosystem of Mt. Ohdaigahara* (46–59), Hadano, Kanagawa: Tokai University Press.

Ando, M., Yokota, H. & Shibata, E. (2003). Bark stripping preference of sika deer, *Cervus nippon*, in terms of bark chemical contents. *Forest Ecology and Management*, *177*, 323–331.

Baldwin, L. K. & Bradfield, G. E. (2007). Bryophyte responses to fragmentation in temperate coastal rainforests: a functional group approach. *Biological Conservation*, *136*, 408–422.

Bates, J. W. (1998). Is "life-form" a useful concept in bryophyte ecology?, *Oikos*, *82*, 223–237.

Doei, H. (1988). Bryophytes of Mt. Ohdaigahara in the Kii Peninsula, Central Honshu, Japan, I. *Nanki Biology*, *30*, 14–23.

Ellyson, W. J. T. & Sillett, S. C. (2003). Epiphyte communities on Sitka spruce in an old-growth redwood forest. *The Bryologist*, *106*, 197–211.

Franks, A. J. & Bergstrom, D. M. (2000). Corticolous bryophytes in microphyll fern forests of south-east Queensland: distribution on Antarctic beech (*Nothofagus moorei*). *Austral Ecology*, *25*, 386–393.

Fritz, Ö. (2009). Vertical distribution of epiphytic bryophytes and lichens emphasizes the importance of old beeches in conservation. *Biodiversity and Conservation*, *18*, 289–304.

Fuertes, E., Burgaz, A. R. & Escudero, A. (1996). Pre-climax epiphyte communities of bryophytes and lichens in Mediterranean forests from the Central Plateau (Spain). *Vegetatio*, *123*, 139–151.

Giordano, S., Sorbo, S., Adamo, P., Basile, A., Spagnuolo, V. & Cobianchi, R. C. (2004). Biodiversity and trace element content of epiphytic bryophytes in urban and extra urban sites of southern Italy. *Plant Ecology*, *170*, 1–14.

Hale, M. E. (1952). Vertical distribution of cryptogams in a virgin forest of Wisconsin. *Ecology*, *33*, 398–406.

Hazell, P., Kellner, O., Rydin, H. & Gustafsson, L. (1998). Presence and abundance of four epiphytic bryophytes in relation to density of aspen (*Populus tremula*) and other stand characteristics. *Forest Ecology and Management*, *107*, 147–158.

Holz, I. & Gradstein, R. S. (2005). Cryptogamic epiphytes in primary and recovering upper montane oak forests of Costa Rica–Species richness, community composition and ecology. *Plant Ecology*, *178*, 89–109.

Hosokawa, T. & Odan, N. (1957). The daily compensation period and vertical ranges of epiphytes in a beech forest. *Journal of Ecology*, *45*, 901–905.

Hylander, K., Nilsson, C., Jonsson, B. G. & Göthner, T. (2005). Substrate form determines the fate of bryophytes in riparian buffer strips. *Ecological Applications*, *15*, 674–688.

Iwatsuki, Z. (1960). The epiphytic bryophyte communities in Japan. *Journal of Hattri Botanical Laboratory*, *22*, 159–351.

Iwatsuki, Z. (2001). *Mosses and liverworts of Japan*. Chiyodaku, Tokyo: Heibonsha.

Kenkel, N. C. & Bradfield, G. E. (1981). Ordination of epiphytic bryophyte communities in a wet-temperate coniferous forest, south-coastal British Columbia. *Vegetatio*, *45*, 147–154.

Lara, F. & Mazimpaka, V. (1998). Succession of epiphytic bryophytes in a *Quercus pyrenaica* forest from the Spanish Central Range (Iberian Peninsula). *Hedwigia*, *67*, 125–138.

Lyons, B. Nadkarni, N. M. & North, M. P. (2000). Spatial distribution and succession of epiphytes on *Tsuga heterophylla* (western hemlock) in an old-growth Douglas-fir forest. *Canadian Journal of Botany*, *78*, 957–968.

McCune, B., Amsberry, K. A, Camacho, F. J, Clery, S., Cole, C., Emerson, C., Felder, G., French, P., Greene, D., Harris, R., Hutten, M., Larson, B., Lesko, M., Majors, S., Markwell, T., Parker, G. G., Pendergrass, K., Peterson, E. B., Peterson, E. T., Platt, J., Proctor, J., Rambo, T., Rosso, A., Shaw, D., Turner, R. & Widmer, M. (1997). Vertical profile of epiphytes in a Pacific Northwest old-growth forest. *Northwest Science*, *71*, 145–152.

McCune, B. & Mefford, M. J. (2006). *PC-ORD. Multivariate analysis of ecological data* (Version 5.10). Gleneden Beach, Oregon: MjM Software.

Mežaka, A. & Znotiņa, V. (2006). Epiphytic bryophytes in old growth forests of slopes, screes and ravines in north-west Latvia. *Acta Universitatis Latviensis*, *710*, 103–116.

Mueller-Dombois, D. (1988). Forest decline and dieback - A global ecological problem. *Trends in Ecology and Evolution*, *3*, 310–312.

Mueller-Dombois, D. & Ellenberg, H. (1974). *Aims and methods of vegetation ecology*. Hoboken, New Jersey: John Wiley and Sons.

Nara Local Meteorological Observatory. (1997). *Climate of Nara Prefecture in the last 100 years*. Kitaku, Tokyo: National Printing Bureau.

Nadkarni, N. M. (1984). Epiphyte biomass and nutrient capital of a neotropical elfin forest. *Biotropica, 16*, 249–256.

Oishi, Y. (2011). Protective management of trees against debarking by deer negatively impacts bryophyte diversity. *Biodiversity and Conservation, 20*, 2527–2536.

Oishi, Y. (2012). Decreased epiphytic bryophyte diversity on Mt. Odaigahara, Japan: Causes and implications. In: G. A. Lameed (Ed.), *Biodiversity Enrichment in a Diverse World* (153–170), Rijeka, Croatia: Intech.

Oishi, Y. (2013). Impact of forest decline on epiphytic bryophyte diversity on Mt. Odaigahara. *The Journal of Information Science, 41*, 47–52.

Ojala, E., Mönkkönen, M. & Inkeröinen, J. (2000). Epiphytic bryophytes on European aspen Populus tremula in old-growth forests in northeastern Finland and in adjacent sites in Russia. *Canadian Journal of Botany, 78*, 529–536.

Pereira Alvarenga, L. D. & Pôrto, K. C. (2007). Patch size and isolation effects on epiphytic and epiphyllous bryophytes in the fragmented Brazilian Atlantic forest. *Biological Conservation, 134*, 415–427.

Pharo, E. J. & Zartman, C. E. (2007). Bryophytes in a changing landscape: The hierarchical effects of habitat fragmentation on ecological and evolutionary processes. *Biological Conservation, 135*, 315–325.

Pypker, T. G., Unsworth, M. H. & Bond, B. H. (2006). The role of epiphytes in rainfall interception by forests in the Pacific Northwest II. Field measurements at the branch and canopy scale. *Canadian Journal of Forest Research, 36*, 819–832.

Sillett, S. C. & Rambo, T. R. (2000). Vertical distribution of dominant epiphytes in Douglas-fir forests of the central Oregon Cascades, *Northwest Science, 74*, 44–49.

Snäll, T., Hagström, A., Rudolphi, J. & Rydin, H. (2004). Distribution pattern of the epiphyte *Neckera pennata* on three spatial scales - importance of past landscape structure, connectivity and local conditions. *Ecography, 27*, 757–766.

Sporn, S. G., Bos, M. M., Kessler, M. & Gradstein, S. R. (2010). Vertical distribution of epiphytic bryophytes in an. Indonesia rainforest. *Biodiversity and Conservation, 19*, 745–760.

Yokoyama, S., Maeji, I., Ueda, T., Ando, M. & Shibata, E. (2001). Impact of bark stripping by sika deer, *Cervus nippon*, on subalpine coniferous forests in central Japan. Forest *Ecology and Management*, *140*, 93–99.

In: Moss
Editor: Jorma Mohamed

ISBN: 978-1-63117-396-7

Chapter 8

FORENSIC BOTANY: THE USE OF MOSSES IN CRIMINAL INVESTIGATIONS

***Marco Caccianiga*[1], *Federica Conforti*[2] and *Massimo Lancia*[3]**

[1]Dipartimento di Bioscienze, Università degli Studi di Milano
[2]Scuola di Specializzazione in Medicina Legale, Italy
Università degli Studi di Perugia, Italy
[3]Sezione di Medicina Legale, Università degli Studi di Perugia, Italy

ABSTRACT

Forensic Botany is the application of botanic knowledge to criminal investigations. It is a relatively new science and it includes several subdisciplines, such as Palynology, Dendrochronology, Limnology, Systematic, Ecology and Molecular Biology. Plant material is everywhere and it is frequently found in association with forensic cases, especially outdoor. Nevertheless its potential applications are often unknown or underestimated by the investigators and prosecutors, so botanical traces are usually overlooked. Instead a systematic analysis of botanical evidence can provide useful information about manner and time of death, find out and dating clandestine graves, identify the primary crime scene and help discovering the offender. Botanical evidence is mainly used in forensic field to provide linkages between crime scenes, individuals (victims or suspects) and objects through morphological, anatomic and genetic study of pollens and spores, seeds, living or decayed leaves, plant fragments or algae. Mosses are particularly useful for this purpose. There are over 10.000 species of mosses worldwide; they are ubiquitous (even if they prefer humid and shadowy zones) but their growth substrates and environmental conditions are variable; so they represent good "biological indicators", more reliable than other plant organisms for geolocation. Mosses can grow directly associated to human remains or objects; moss fragments are resistant to decay and they can heavily attach to clothes or shoes, where they can be discovered and identified by genetic analysis even after a long time. Moss spores found in evidentiary materials or human remains might be used to establish in which season a crime occurred. In the last years the use of mosses for forensic purposes expanded; in recent studies the regular growth rate of some species of mosses was used in estimating the minimum post-mortem interval of skeletal remains. In this chapter the authors give an overlook on possible application of moss in Forensics.

INTRODUCTION

Forensic Botany is the application of botanic knowledge to criminal investigations. The role of plant science in such investigation has proved more and more useful in the recent decades [1] [2]; many botanical disciplines have been involved such as palynology [3], macroremains analysis, dendrochronology [4] [5], study of vegetation dynamics [6]. Plant material is widespread and is frequently found in association with forensic cases, especially outdoor. Nevertheless its potential applications are often unknown or underestimated by the investigators and prosecutors, and botanical traces are usually overlooked. The main questions addressed by forensic botany consist of spatial and temporal reconstructions, such as identification of previous geographical or ecological location of remains and individuation of the primary crime scene, identification of clandestine burials, estimation of post-mortem interval (PMI - time from when death occurs until the body is discovered). Plant fragments (seeds, fruits, flowers, leaves and other vegetative parts) are a powerful tool to link a body or any other evidence back to a primary crime scene as they can provide detailed information about the ecological and geographic location where the body has been exposed; their primary importance has been outlined in forensic-like investigations carried out in archeological context [7]. Living plants can add further information, particularly for dating surfaces or remains and for recording events which may have occurred within a site affecting plant growth, like passage of vehicles, excavation of graves, trampling, etc. Bryophytes are particularly useful for all these purposes.

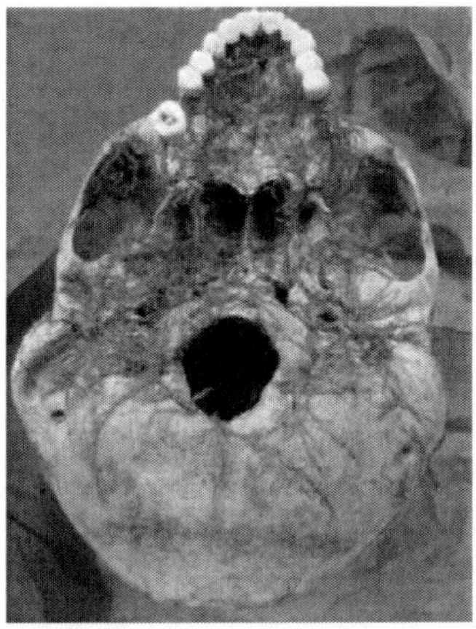

Figure 1. Mosses growing on the lower part of a skull.

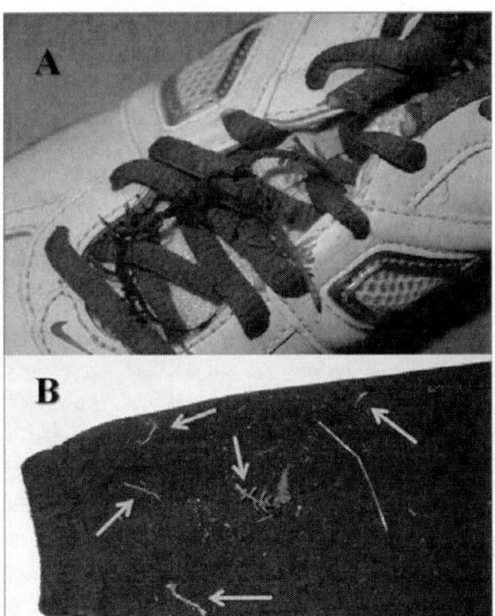

Figure 2. A: moss fragments attached to a shoe; B: plant and moss (arrows) fragments attached to a sweater.

They are ubiquitous (even if they prefer humid and shadowy zones) but at the same time they are strongly influenced by substrate and site condition at local scale, thus representing good biological indicators. Bryophytes can grow directly associated to human remains or objects (Figure1); moss fragments are resistant to decay and they can attach to clothes or shoes (Figure 2). Unfortunately, the use of bryophytes in forensic research has been uncommon so far, well below the potentiality of this kind of evidence.

In this chapter the authors give an overlook on possible application of bryophytes within forensic sciences.

Bryophytes As Ecological Indicators

Plants are known to be powerful ecological indicators, as their occurrence is the visible synthesis of underlying environmental factors, which are often invisible and/or difficult to measure. The association of plants or plant fragments to human remains or other objects of forensic interest could thus provide information about their present and past location. Within this context, methodologies close to those of paleobotany are often involved to investigate the occurrence of plant remains, their identification and their interpretation. Bryophytes are not an exception, and their occurrence could add further detail to the information provided by vascular plant assemblages. As previously stated, the peculiarities of bryophytes lie in their strict association with the findings, the probability of collecting them as living, complete individuals rather than partial fragments and their sensitivity to microscale environmental variability. In forest ecosystems, bryophytes richness has been shown to depend from local canopy cover of tree species, litter cover and, in case of species growing on boulders, boulder size [8]. Such detailed information has proved useful in some forensic cases such as that reported by Caccianiga and Compostella [9], in which the previous spatial position of a human skull found on a tree had to be reconstructed.

In order to fully exploit their potential of information, the availability of detailed ecological databases for bryophyte species, such as that of Hill et al. [10], is a critical issue. Fortunately, also texts aimed to moss identification often give important ecological indications as, for Italian moss flora, that by Cortini Pedrotti [11]. A classification of life strategies based on few life history traits as that proposed by During [12] may help to relate some macroscopic, consistent syndromes with the main environmental traits.

BRYOPHYTES GROWTH AND POST-MORTEM INTERVAL

The occurrence of living plants associated with objects of forensic interest (usually human remains, but also artifacts, surfaces, fillings of suspected clandestine graves, etc) may provide important indications about the minimum age of permanence of such objects in the present location and position. Dendrochronology (the analysis of tree growth rings) represents a powerful tool, particularly when applied to roots or stems directly growing on or within remains (Figures 3-4).

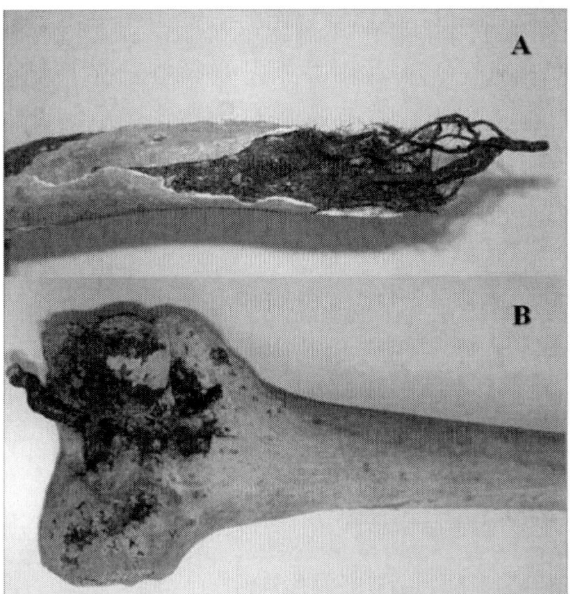

Figure 3. Roots directly growing on and within skeletal remains (A: right clavicle; B: distal epiphysis of the left femur).

On the other hand, a direct association of woody species to remains or objects of forensic interest is not always possible, particularly when remains lie above ground. In this case, lower plants (algae, lichens, mosses and liverworts) growing on the findings could give a minimum time span of persistence, provided that their age could be assessed with sufficient accuracy.

In mosses, discrete growth units can be recognized on the basis of morphological criteria [13] [14]; these units have been shown to correspond to annual increments [14] [15] [16] thus giving the opportunity to evaluate the age of moss individuals growing in the investigated surface.

Figure 4. Cross sections of roots collected on human remains prepared for dendrochronological analysis.

This method was applied successfully for forensic purposes on human remains [9] [17] [18] (Figures 5-6).

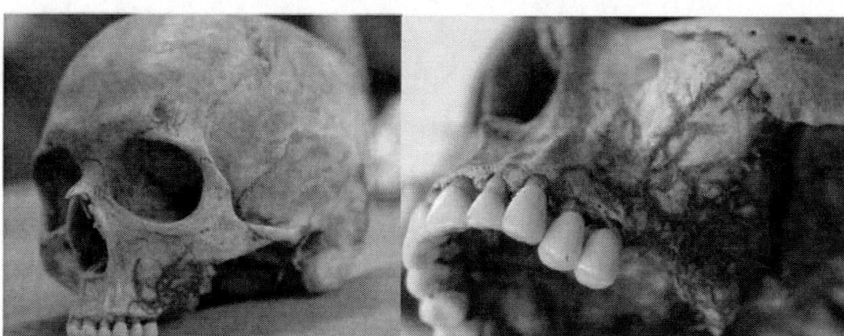

Figure 5. Moss growing on left maxillar bone.

All the cited case studies involved mosses growing on skeletal remains (skull and tibia) and clothes (only in Cardoso et al., 2010) and outlined a minimum post-mortem interval ranging from 2 (Lancia et al., 2013) to 6 (Caccianiga and Compostella, 2009) years. In two cases, the positive identification of the remains could give a check of the estimation. In the case reported by Lancia et al., the estimation was of 24-30 months and the person disappeared 40 months earlier; in that reported by Cardoso et al. the disappearance of the presumptive individual occurred 6 years before and the estimation gave a result of 3 years. Such discrepancy is expected, as the estimation always gives minimum interval: the time of decomposition and that required by propagules to reach the remains must be added.

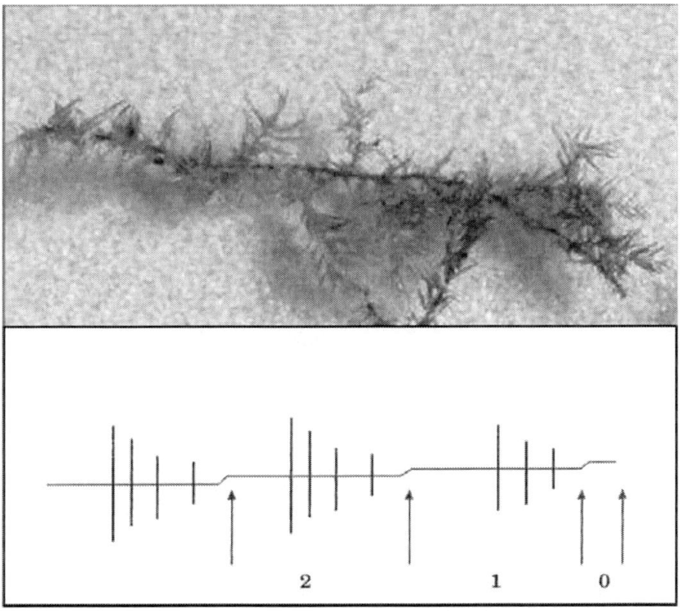

Figure 6. Estimation of moss age: comparison between growth units of a moss associated with skeletal remains and regular monopodial growth pattern.

Both these parameters could be very variable: decomposition is strongly affected by environmental factors such as temperature, moisture, insect and faunal activity, occurrence of clothes, etc; the time of arrival of spores and vegetative propagules is affected by species-specific dispersal ability, distance of parent individuals, microtopographic and microclimatic features. A certain time lag between the exposure of bones and moss colonization seems to be the rule, as skeletal remains recently exposed (one-two years or less) seldom show a moss cover, even in under forest canopy. On the other hand, favorable environment (e.g. high humidity) can enhance moss growth resulting in high cover after a relatively short time, as shown by Lancia et al. [18]. (Figure 7).

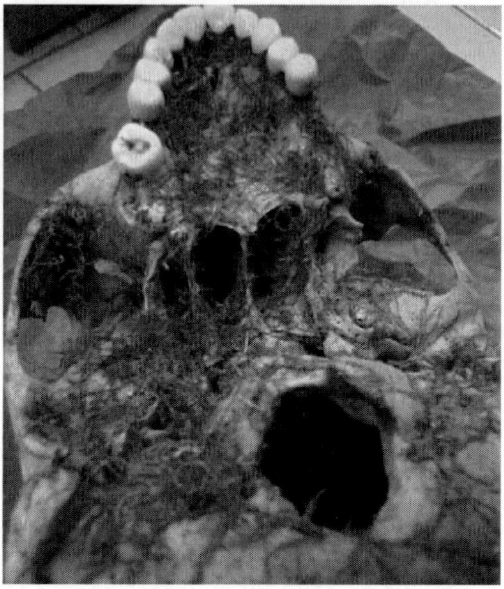

Figure 7. High moss colonization in the lower part of a skull, in the case reported by Lancia et al. (2013).

A further information to be kept in mind in the evaluation of the minimum age of a moss individual is the life expectancy of the shoots, in order to evaluate the extent of the loss on the trailing edge of the growing shoot. For some species such as *Polytrichum commune* shoots are reported to remain active for 3-4 years [12]; dead shoots may be used to trace back the life of a plant for a long time, up to 50 years for a cushion of *Leucobryum glaucum* [12].

Within this frame, some caveats must be kept in mind. Moss growth does not always follow a yearly pattern but can be strongly influenced by water availability, with pulses occurring in different periods during the year: Rincon and Grime [19] observed rapid growth rates in autumn for *Brachytecium rutabulum,* by which the species is able to overgrow the accumulating litter, while *Hypnum cupressiforme* and *Isothecium myosuroides* were observed to perform high growth rates in late autumn-winter [20]. Hanslin [16] reported variable growth rates during the year for some boreal forest species, with some of them (e.g. *Ptilium crista-castrensis*) particularly affected by drought. For this reason, at least for some species, more growth units may occur for a single year, severely affecting age estimation. The knowledge of the specific ecological traits of the investigated species is thus, again, of great importance in order to avoid dangerous mistakes.

Occurrence and Analysis of Bryophytes Macroremains

All the described evidence could be obtained by bryophyte occurrence provided that their fragment could get attached to the investigated objects or remains, preserved through time and, finally, recognized. Apart from living individuals actively growing on the investigated surfaces, the occurrence of bryophyte fragments could be easily overlooked, given their small dimension which can easily shrink under dry environmental conditions. However, moss leaves can be found in good quantity because of their ability to adhere to clothes and shoes. Virtanen et al. [21] showed experimentally that moss fragments were able to adhere to boots of people walking outdoors and remained attached even after several hours of walking on dry environment such as roads.

The identification of bryophytes species is not easy, particularly when performed on small and incomplete samples such as leaflets or small fragments in general. However, many diagnostic traits for the identification lie in leaf features, either macroscopic (shape, dimension) or microscopic (cell shape, nerves, occurrence of alar cell, leaf blade margin, etc), which may allow the identification in many cases of incomplete specimens.

DNA-based identification could overcome many difficulties occurring when morphological evidence is scarce; the clonal origin of many bryophyte population may allow very detailed responses. Korpelainen and Virtanen [22] reported a homicide case in which small shoot particles of a bryophyte species found in the suspect's car were demonstrated to originate from the crime site in a forest. Virtanen et al. [21] experimentally demonstrated that moss DNA could remain preserved enough to allow fingerprinting after remaining for 18 months in paper bags at temperature ranging from -28° to +25°C.

Such evidence, although still at the beginning, clearly indicates the potentiality of the use of moss fragments for forensic reconstruction, even if the development of sets of microsatellite markers to allow genetic profiling for a wide range of bryophyte species is of decisive importance.

CONCLUSION

This short overview clearly demonstrates that mosses and relatives should surely be taken into account together with vascular plants within the growing field of forensic botany. The morphological, ecological and biological peculiarities of bryophytes make their evidence complementary to those provided by vascular plants. Unfortunately, taxonomic skill and ecological knowledge are often difficult to find for investigators and forensic bryology plays a still minor role within the already underexplored field of forensic botany. The availability of reference collections for the identification of incomplete specimens, of ecological databases and of molecular information is a major point to make possible the exploration of the potentiality of such evidence.

REFERENCES

[1] Miller Coyle, H.; Ladd, C.; Palmbach, T.; Lee, H.C. (2001). The Green Revolution: Botanical Contributions to Forensics and Drug Enforcement. *Croatian Medical Journal*, 42 (3), 340-345.

[2] Miller Coyle, H.; Lee, C.L.; Lin, W.Y.; Lee, H.C.; Palmbach, T.M. (2005). Forensic Botany: Using Plant Evidence to Aid in Forensic Death Investigation. *Croatian Medical Journal*, 46 (4), 606-612.

[3] Mildenhall, D.C. (2006). Hypericum pollen determines the presence of burglars at the scene of crime: an example of forensic palynology. *Forensic Science International*, 163, 231-235.

[4] Denne, M.P. (1977). Dating of events from tree growth and wood structure. *Journal of Forensic Sciences Society*, 17, 257-264.

[5] Willey, P.; Heilman, A. (1987). Estimating time since death using plant roots and stems. *Journal of Forensic Sciences*, 32, 1264-1270.

[6] Caccianiga, M.; Bottacin, S.; Cattaneo, C. (2012). Vegetation Dynamics as a Tool for Detecting Clandestine Graves. *Journal of Forensic Sciences,* 57 (4), 983-988.

[7] Heiss, A.G.; Oeggl, K. (2009). The plant macro-remains from the Iceman site (Tisenjoch, Italian-Austrian border, eastern Alps): new results on the glacier mummy's environment. *Vegetation History & Archaeobotany*, 18: 23 - 35.

[8] Weibull, H; Rydin, H. (2005). Bryophyte species richness on boulders: relationship to area, habitat diversity and canopy tree species. *Biological Conservation*, 122, 71–79.

[9] Caccianiga, M; Compostella, C (2009). Botanica e zoologia forense. In: Intini A, Picozzi M editors. Scienze Forensi: teoria e prassi dell'investigazione scientifica. Torino: UTET; 2009; 230-257.

[10] Hill, MO; Preston, CD; Bosanquet, SDS; Roy, DB (2007). BRYOATT- Attributes of British and Irish Mosses, Liverworts and Hornworts With Information on Native Status, Size, Life Form, Life History, Geography and Habitat. NERC Centre for Ecology and Hydrology and Countryside Council for Wales.

[11] Cortini Pedrotti, C. Flora dei muschi d'Italia. Roma, IT: Antonio Delfino Editore; 2001.

[12] During, H.J. (1979). Life strategies of Bryophytes: a preliminary review. *Lindbergia*, 5, 2-18.

[13] Potter, J.A.; Press, M.C.; Callaghan T.V.; Lee, J.A. (1995). Growth responses of Polytrichum commune and Hylocomium splendens to simulated environmental change in the sub-arctic. *New Phvtologist*, 131 (4), 533-541.

[14] Zechmeister, H. G. (1995). Growth Rates of Five Pleurocarpous Moss Species Under Various Climatic Conditions. *Journal of Bryology*, 18, 455–468.

[15] Bisang, I.; Ehrlén, J; Hedenäs, L. (2008). Are Annual Growth Intervals Independent Units in The Moss Pseudocalliergon Trifarium (Amblystegiaceae). *The Bryologist*, 111 (3), 435-443.

[16] Hanslin, H.M. (1999). Seasonal dynamics of biomass increase and shoot elongation in five co-occurring boreal forest bryophytes. *Journal of Bryology,* 21, 5-15.

[17] Cardoso, H.F.; Santos, A.; Dias, R.; Garcia, C.; Pinto, M.; Sérgio, C.; Magalhães, T. (2010). Establishing a minimum postmortem interval of human remains in an advanced state of skeletonization using the growth rate of bryophytes and plant roots. *International Journal of Legal Medicine*, 124 (5), 451-456.

[18] Lancia, M; Conforti, F.; Aleffi, M; Caccianiga, M.; Bacci, M; Rossi, R. (2013). The use of Leptodyctium riparium (Hedw.) Warnst in the estimation of minimum postmortem interval. *Journal of Forensic Sciences*, 58, 239-242.

[19] Rincon, E.; Grime, J.P. (1989). An analysis of seasonal pattern of bryophyte growth in a natural habitat. *Journal of Ecology*, 77, 447–55.

[20] Pitkin, P.H. (1975). Variability and seasonality of the growth of some corticolous pleurocarpous mosses. *Journal of Bryology*, 8, 337-356.

[21] Virtanen, V.; Korpelainen, H.; Kostamo, K. (2007). Forensic botany: Usability of bryophyte material in forensic studies. *Forensic Science International,* 172, 161–163.

[22] Korpelainen, H.; Virtanen, V. (2003). DNA fingerprinting of Mosses. *Journal of Forensic Sciences*, 48, 804–807.

INDEX

D

E

F

G

H

I

J

K

L

M

N

O

P

Q

R

S

T

U

V

W

Y

Z